IEEE Radio and Antenna Days of the Indian Ocean (IEEE RADIO 2015)

IOP Conference Series: Materials Science and Engineering Volume 120

Belle Mare, Mauritius
21 – 24 September 2015

Editor:

Vikass Monebhurran

ISBN: 978-1-5108-2411-9
ISSN: 1757-8981

Printed from e-media with permission by:

Curran Associates, Inc.
57 Morehouse Lane
Red Hook, NY 12571

CURRAN ASSOCIATES INC. proceedings.com

Some format issues inherent in the e-media version may also appear in this print version.

Printed by Curran Associates, Inc. (2016)

Institute of Physics
Dirac House, Temple Back
Bristol BS1 6BE UK

Phone: 44 1 17 929 7481
Fax: 44 1 17 920 0979

techtracking@iop.org

Additional copies of this publication are available from:

Curran Associates, Inc.
57 Morehouse Lane
Red Hook, NY 12571 USA
Phone: 845-758-0400
Fax: 845-758-2633
Email: curran@proceedings.com
Web: www.proceedings.com

TABLE OF CONTENTS

RADIO FREQUENCY MAPPING USING AN AUTONOMOUS ROBOT: APPLICATION TO THE 2.4 GHZ BAND 1
J M Lebreton, N M Murad, R Lorion

A REAL-TIME SPECTRUM HANDOFF ALGORITHM FOR VOIP BASED COGNITIVE RADIO NETWORKS: DESIGN AND PERFORMANCE ANALYSIS 5
Tamal Chakraborty, Iti Saha Misra

CLUSTER-HEAD TECHNIQUES FOR SINGLE-HOP ROUTING PROTOCOL IN ENERGY EFFICIENT WIRELESS SENSOR NETWORKS 9
Somasekhar Kandukuri, Nour Murad, Richard Lorion

A LOW-COST RSSI-BASED LOCALIZATION SYSTEM FOR WILDLIFE TRACKING 13
Markus Hartmann, Thorsten Nowak, Lucila Patino-Studencki, Jörg Robert, Albert Heuberger, Jörn Thielecke

LOW-COST INKJET ANTENNAS FOR RFID APPLICATIONS 17
T Çiftçi, B Karaosmanoglu, Ö Ergül

RF PULSE SIGNAL INTEGRITY ANALYSIS FOR NONLINEAR ENDED MICROSTRIP LINE ATOM-PROBE TOMOGRAPHY 21
L. Zhao, A. Delamare, A. Normand, F. Delaroche, O. Latry, F. Vurpillot, B. Ravelo

COMPACT WIDE STOPBAND LOWPASS FILTER WITH HIGH SUPPRESSION USING CORRUGATED TRANSMISSION LINE 25
P M Raphika, P Abdulla, P M Jasmine

DC-PASS FILTER DESIGN WITH NOTCH FILTERS SUPERPOSITION FOR CPW RECTENNA AT LOW POWER LEVEL 29
J Rivière, A Douyère, F Alicalapa, J-D Lan Sun Luk

SIMULATIONS AND TESTS OF PROTOTYPE ANTENNA SYSTEM FOR LOW FREQUENCY RADIO EXPERIMENT (LORE) SPACE PAYLOAD FOR SPACE WEATHER OBSERVATIONS 33
Kaiwalya Pethe, Shridhar Galande, Sachin Jamadar, S. P. Mahajan, R. A. Patil, B. C. Joshi, P. K. Manoharan, Jayashree Roy, G. Kate

HYPERBAND BI-CONICAL ANTENNA DESIGN USING 3D PRINTING TECHNIQUE 37
J. A. Andriambeloson, P. G. Wiid

INVESTIGATION OF SPHERICAL AND CYLINDRICAL LUNEBURG LENS ANTENNAS BY THE GREEN'S FUNCTION METHOD 41
S. Knyazev, A. Korotkov, B. Panchenko, S. Shabunin

MITRA 2.0 45
Vinand Prayag, Nazir Vydelingum, Girish Kumar Beeharry

SWAMURAY - SWAPPING MEMORY UNIT FOR RADIO ASTRONOMY 49
Simon Winberg

LOW FREQUENCY RADIO EXPERIMENT (LORE) 53
P. K. Manoharan, Arun Naidu, B. C. Joshi, Jayashree Roy, G. Kate, Kaiwalya Pethe, Shridhar Galande, Sachin Jamadar, S. P. Mahajan, R. A. Patil

STATISTICAL PERFORMANCES OF RESISTIVE ACTIVE POWER SPLITTER 57
Sébastien Lalléchère, Blaise Ravelo, Atul Thakur

MODELLING OF POLARIZATION ROTATOR OF THE BROADBAND SCANNING ANTENNA ARRAY 61
A I Semenikhin, D V Semenikhina, Y V Yukhanov

CONDUCTOR DISC USED TO SUPPRESS SPURIOUS MODE AND ENHANCE ELECTRIC COUPLING IN A DIELECTRIC LOADED COMBLINE RESONATOR 65
T M Pholele, J M Chuma

IMPACT OF ELECTROMAGNETIC FIELDS ON HUMAN VESTIBULAR SYSTEM AND STANDING BALANCE: PILOT RESULTS AND FUTURE DEVELOPMENTS 69
A Allen, S Villard, M Corbacio, D Goulet, M Plante, M Souques, F Deschamps, G Ostiguy, J Lambrozo, A W Thomas, A Legros

Author Index

IOP Conf. Series: Materials Science and Engineering **120** (2016) 011001 doi:10.1088/1757-899X/120/1/011001

EDITOR'S FOREWORD

Her Excellency, Dr. Ameenah Gurib-Fakim, President of the Republic of Mauritius, officially opened the third edition of the Radio and Antenna Days of the Indian Ocean (IEEE RADIO 2015) that was held at the Long Beach Resort, Belle Mare, Mauritius from 21st to 24th September 2015. It was an honor and a great pleasure for all those involved in its organization to welcome the participants to IEEE RADIO 2015.

The aim of the conference is to discuss recent developments, theories and practical applications covering the whole scope of radio-frequency engineering, including radio waves, antennas, propagation, and electromagnetic compatibility. The Organizing Committee consisted of scientists from CentraleSupélec, France and the University of Mauritius. All of them are also active members of the Radio Society (Mauritius) that organized IEEE RADIO 2015 in collaboration with the IEEE Antennas and Propagation Society. Union Radio Scientifique Internationale (URSI) provided technical co-sponsorship for the conference. We are thankful to CST, a major player in computational electromagnetics, for financially supporting this event for the third time. Following numerous requests, the Island of Mauritius, worldwide known for its white sandy beaches and pleasant tropical atmosphere, was once again chosen for the organization of the 3rd edition of the RADIO international conference.

IEEE RADIO 2015 brought together participants from about twenty different countries. It featured sixteen oral sessions on state-of-the-art research themes including antenna design, electromagnetic compatibility, wireless communication, wireless power transfer, numerical modeling and medical and industrial applications of electromagnetic fields. A half-day workshop was organized by CST on the first day of the conference to present the latest version of the CST Studio Suite. Three internationally recognized scientists delivered invited lectures. All accepted papers that were effectively presented during the conference were submitted to the IEEE Xplore Digital Library.

In line with the previous editions, a call to extended contributions for publication as a volume in the *IOP Conference Series: Materials Science and Engineering* (MSE) was made. Following the call, thirty full papers were received. All submitted contributions were then peer-reviewed, revised according to the reviewer's comments, and eventually accepted or rejected. Eighteen papers were ultimately accepted for publication in this volume.

It is believed that the IOP Conference Series dedicated to the RADIO international conference provides an excellent source of pertinent and long-lasting scientific material in the numerous fields covered by conference.

Editor
Vikass Monebhurrun

Editorial Board Members

IOP Conf. Series: Materials Science and Engineering **120** (2016) 011003 doi:10.1088/1757-899X/120/1/011003

Peer review statement

All papers published in this volume of *IOP Conference Series: Materials Science and Engineering* have been peer reviewed through processes administered by the proceedings Editors. Reviews were conducted by expert referees to the professional and scientific standards expected of a proceedings journal published by IOP Publishing.

IOP Conf. Series: Materials Science and Engineering **120** (2016) 012001 doi:10.1088/1757-899X/120/1/012001

Radio Frequency Mapping using an Autonomous Robot: Application to the 2.4 GHz Band

J M Lebreton, N M Murad and R Lorion

LE2P Laboratory, University of La Reunion, Saint-Pierre, France

E-mail: jean.lebreton@univ-reunion.fr

Abstract. Radio signal strength measurement systems are essential to build a Radio Frequency (RF) mapping in indoor and outdoor environments for different application scenarios. This paper presents an autonomous robot making the construction of a radio signal mapping, by collecting and forwarding different useful information related to all access point devices and inherent to the robot towards the base station. A real case scenario is considered by measuring the RF field from our department network. The RF signal mapping consistency is shown by fitting the measurements with the radio signal strength model in two-dimensional area, and a path-loss exponent of 2.3 is estimated for the open corridor environment.

1. Introduction

Radio Frequency (RF) mapping is generally performed in wireless site surveys in order to monitor the coverage of a wireless network. Systems and methods for surveying a wireless network site are presented in [1]. The RF coverage is generally based on measurements taken by a mobile human user. Measurement consists of repetitive tasks which are time consuming and can lead to measurement errors regarding the coordinate information. Other measurement campaigns are done with sensor nodes placed on transport vehicles such as [2], however this method does not enable a controllable path by the sensor node itself.

An autonomous mobile robot is proposed as a measurement technique to avoid these drawbacks. Measurements are also taken for different positions since the robot can move inside an environment and record its path. The data are transferred in real-time to the base station using a wireless transmission. Then the robot actual location and the data exploitation can be directly observed by the user. The advantages of this method are the fact that measurements are performed in an autonomous way and the mapping is done in real-time. The user's presence on the site is only needed at the beginning of the measurement process.

Furthermore, RF mapping can be used to find and identify the different RF sources available in the environment. This automated system can then be exploited for other projects which aim at harvesting RF energy. Jabbar presents the overview and progress achieved in RF energy harvesting field in [3]. The amount of harvested energy is still low, so it is important to know where the most amount of energy is available. The automated system can also produce RSSI (Received Signal Strength Indicator) fingerprinting of an area, which is used for WiFi localisation in some studies [4],[5]. Our main contribution is the automation of a reliable RF mapping by a robot with the RF distribution displayed in real-time.

IOP Conf. Series: Materials Science and Engineering **120** (2016) 012001 doi:10.1088/1757-899X/120/1/012001

2. Robot Hardware

The robot, illustrated in figure 1, is based on the BOE Shield-Bot from Parallax Inc. The BOE Shield-Bot kit is actually composed of a Board of Education (BOE) Shield, a metal chassis, servo motors and wheels. An Arduino Uno is plugged in underneath the Shield. Arduino is an open-source electronics platform based on easy-to-use hardware and software. The Arduino Uno is a micro-controller board that controls all the components connected to the robot. An ultrasonic distance sensor is mounted at the front of the robot so that it can detect obstacles. Then, a gyroscope is fixed to the chassis in order to measure the angular velocity of the robot during its rotation. Concerning the wireless communication, two kinds of modules are used. The Arduino WiFi shield is used to measure signal strength from the different access points. The other wireless transmission is through an Xbee module mounted on the Xbee shield, providing data transmission from the robot to the base station. The robot has a size of 120x160x130 mm and a weight of 0.6 kg. Finally, the batteries allow the robot to work more than 4 hours, corresponding to a 2 km navigation.

3. Principle and Algorithm

The RF mapping principle is to let the robot move within a defined area and to frequently make a scan of the available RF signals. For practical applications, the 2.4 GHz band is selected for the case study. WiFi access points are then considered as RF sources for the study. The robot has to make scans of the available WiFi networks at different positions. The SSID (Service Set Identifier), RSSI and encryption types of the different access points are given by the scans. Even if the 13 channels can be scanned, only the five strongest signals are recorded here in order to keep data in a uniform size. The robot is also constrained to avoid obstacles while it is moving. Random rotations of the robot have been added during the travel in order to have a better coverage of the measured area. The algorithm used by the robot is defined in figure 1.

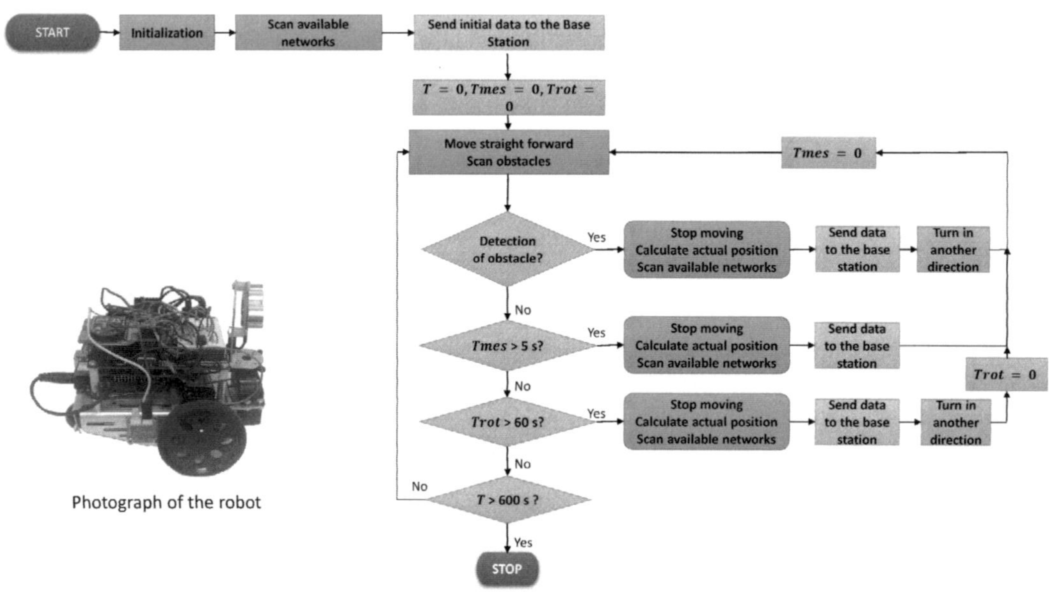

Figure 1. Flowchart of the algorithm used by the robot to perform a RF mapping of an area.

4. Measurement campaign

A RF mapping of our department was performed. The floor plan is illustrated in Figure 2 and the access points locations are represented as well. The network deployment and configuration have been done by the department and no specific change is done for the measurement process in order to study the network state. The results are presented in figure 2, which shows the mapping obtained after 45 minutes of operating time. The robot moved along the corridor of the building and the path is represented in the upper-left sub-figure. At each point, the robot measured the RSSI from five access points since only the five highest signal strengths are considered.

Besides, the robot travelled within an area of 400 m^2. It was located upstairs, so measurements could not be obtained in the middle of the area and signal strengths are only represented in the corridor. The RSSI distribution related to the different sources is represented in figure 2. It shows a reliable mapping for the first three sources, which were the closest access points to the robot path. For all of them, the RSSI distribution is consistent since high signal strengths are observed next to the related access point locations. The same signal strength levels are also obtained from these sources, which is consistent with their location toward the robot path. Then, some data have been collected for two other sources, but they were not close enough to get a reliable map. Therefore, the robot can carry out a good RF mapping of a network in a real case study. The entire mapping process can be viewed in [6].

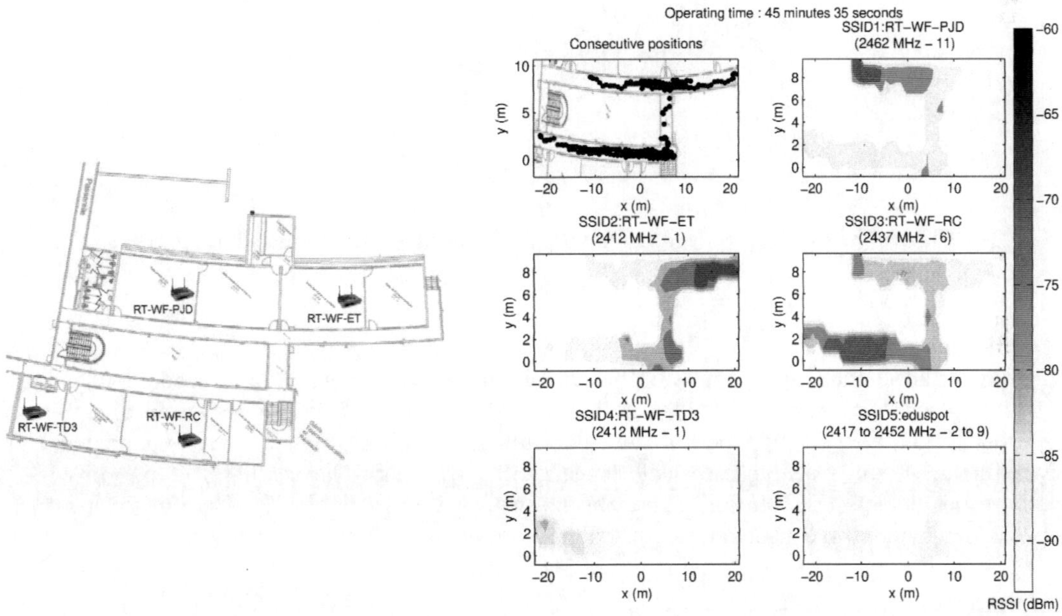

Figure 2. RF mapping of the department obtained after 45 minutes of measurement.

5. Application to the propagation model characterization

The measurements performed by the robot can also be exploited to estimate the parameters of the radio signal propagation, which is studied according to the log-normal path loss model [7]:

$$P_{dBm} = P_0 - 10n \log(||X - X_s||) + Z_\sigma \tag{1}$$

where P_0 is the received power at 1 meter from the source, n is the path loss exponent, X and X_s are respectively the positions of the receiver and transceiver, Z_σ is a normally distributed variable with a standard deviation σ and a zero-mean value ($Z_\sigma \sim \mathcal{N}(0, \sigma)$). Z_σ not only represents the shadowing and multipath fading related to the signal propagation, but also the measurement errors from the device.

A least-square method is applied to fit the measurements with equation (1) in order to estimate P_0 and n. A received power P_0 of -53.6 dBm and a path loss exponent n of 2.3 are estimated, which is relevant for this type of environment. So this new path loss exponent can be referred as the path loss exponent in an open corridor environment. The standard deviation σ is then calculated through equation 1, resulting in $\sigma = 5.7$ dBm. The signal strength attenuation is plotted in figure 3 using the model with estimated parameters or measured data. The measured data fit to the model even though there is a significant noise, which is inherent to a radio signal propagation.

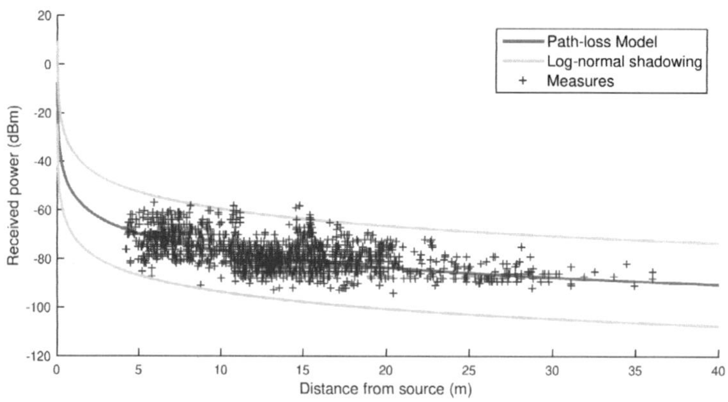

Figure 3. Signal strength attenuation model fitted to the measurements.

6. Conclusion

This paper shows that an accurate RF mapping can be performed by an autonomous robot, avoiding a user to walk around for collecting data. The measured data is also used to estimate the propagation model parameters of the environment, providing the path loss exponent of the environment. As our research project is concerned by RF energy harvesting, the map provided by the robot depicts the location where the most energy is available. It is useful for a future network deployment of sensor nodes with energy harvesting capabilities.

References

[1] Hrastar S E 2009 Systems and methods for wireless network site survey *U.S. Patent* no. 7522908

[2] Hasenfratz D et al. 2013 *Proceedings of First International Workshop on Sensing and Big Data Mining* (New York: ACM) pp. 1-6

[3] Jabbar H, Song Y S and Jeong T T 2010 *IEEE Transaction on Consumer Electronics* 56 pp. 247-253

[4] Biswas J and Veloso M 2010 *IEEE International Conference on Robotics and Automation* (Los Alamitos: IEEE Computer Society) pp. 4379-4384

[5] Husen M N and Lee S 2014 *Proceedings of the 8th International Conference on Ubiquitous Information Management and Communication* (New York: ACM) pp. 109-114

[6] Lebreton J 2014 Real time visualization of the RF mapping in the R&T department (32x speed) https://www.youtube.com/watch?v=zh-WkN14Tpw

[7] Vallet J et al. 2013 10th *IEEE Workshop on Positioning Navigation and Communication* (Los Alamitos: IEEE Computer Society) pp. 1-6

A Real-time Spectrum Handoff Algorithm for VoIP based Cognitive Radio Networks: Design and Performance Analysis

Tamal Chakraborty, Iti Saha Misra

Department of Electronics and Telecommunication Engineering

Jadavpur University, Kolkata 700032, West Bengal, India

tamalchakraborty29@gmail.com

Abstract. Secondary Users (SUs) in a Cognitive Radio Network (CRN) face unpredictable interruptions in transmission due to the random arrival of Primary Users (PUs), leading to spectrum handoff or dropping instances. An efficient spectrum handoff algorithm, thus, becomes one of the indispensable components in CRN, especially for real-time communication like Voice over IP (VoIP). In this regard, this paper investigates the effects of spectrum handoff on the Quality of Service (QoS) for VoIP traffic in CRN, and proposes a real-time spectrum handoff algorithm in two phases. The first phase (VAST-VoIP based Adaptive Sensing and Transmission) adaptively varies the channel sensing and transmission durations to perform intelligent dropping decisions. The second phase (ProReact-Proactive and Reactive Handoff) deploys efficient channel selection mechanisms during spectrum handoff for resuming communication. Extensive performance analysis in analytical and simulation models confirms a decrease in spectrum handoff delay for VoIP SUs by more than 40% and 60%, compared to existing proactive and reactive algorithms, respectively and ensures a minimum 10% reduction in call-dropping probability with respect to the previous works in this domain. The effective SU transmission duration is also maximized under the proposed algorithm, thereby making it suitable for successful VoIP communication.

1. Introduction

Cognitive Radio Networks (CRN) [1] aim to mitigate the problem of spectrum congestion by allowing opportunistic access of the licensed frequency bands by the unlicensed Secondary Users (SUs) in absence of the licensed Primary Users (PUs). In such a network, whenever a PU arrives in its default channel, the corresponding SU must vacate the channel and perform successful spectrum handoff to an available idle channel for resuming transmission. An efficient spectrum handoff algorithm must be in place to reduce the handoff delay and potential interference with the PUs [2]. Thus, spectrum mobility is a significant aspect of CRN, where two crucial aspects gain significance; i) intelligent decision mechanism to drop from the current channel without interrupting PU transmissions, and ii) efficient channel selection algorithm to perform handoff without incurring significant delay.

Existing works have primarily focused on designing proactive [2, 3] and reactive [4] handoff strategies, with the objective of decreasing the dropping probability [5] and interference with the PU [2]. While proactive algorithms efficiently reduce the channel selection time by obtaining a pre-determined target channel sequence (T.C.S.) [4], their primary disadvantage stems from the fact that the channels in the T.C.S. may not be idle at the current time slot especially under varying network conditions. Reactive handoff mechanisms solve this problem by selecting channels on an on-demand fashion, but at the cost of higher delay due to rise in channel selection and channel consensus time [5]. Therefore, it is observed from the literature survey that these algorithms fail to address Quality of Service (QoS) constraints (delay, loss, etc.) of real-time communication like Voice over IP (VoIP) [6].

This paper deals with this problem and proposes a real-time spectrum handoff algorithm to maintain the QoS requirements of VoIP calls. The first phase adopts a novel strategy to take dropping decisions based on adaptive tuning of timing parameters and the second phase performs efficient channel selection in real-time without incurring significant delay and call drop. Overall, the proposed algorithm maximizes the duration of successful communication for VoIP calls.

The paper is organized as follows. The proposed algorithm is explained in detail in Section 2 followed by its mathematical formulation in Section 3 and simulation results in Section 4. Finally, the paper is concluded in Section 5.

2. Proposed Algorithm

The proposed algorithm is comprised of two segments. The first segment, *VAST* (VoIP based Adaptive Sensing and Transmission) deploys adaptive sensing and transmission durations for SUs, and subsequently defines the conditions for dropping from the current channel. This is followed by the execution of the second segment, *ProReact* (Proactive and Reactive Handoff) which performs successful channel handoff to a new idle available channel, without degrading the QoS limits (delay, loss) of VoIP calls. Both these segments are depicted in Figure 1 and are described as follows.

2.1. VAST (VoIP based Adaptive Sensing and Transmission)

The *VAST* segment implements adaptive sensing and transmission durations for SUs, and also defines the conditions for dropping from the current channel. Two time intervals of high and low duration are considered each for channel sensing and transmission (denoted by t_{ls}, t_{hs} for sensing and t_{ld}, t_{hd} for transmission). If the channel is sensed idle during t_{ls}, SU transmits during t_{hd} period. If the channel is busy, SU enters t_{hs} period to re-sense the channel. At the end of every t_{hs}, SU starts transmission in t_{ld} period if the channel is idle and resets its sensing duration to t_{ls} for the next sense cycle.

Based on the four parameters (t_{ls}, t_{hs}, t_{ld}, t_{hd}), the decision to drop from the current channel is taken as per three rules; i) *Cons_sense*: drop from channel if sensed busy consecutively in two t_{hs} intervals, ii) *Cons_tx*: drop from channel if SU performs transmission in two t_{ld} intervals in consecutive cycles, iii) *Alt_tx*: drop from channel if no. of transmissions in t_{ld} is greater than half the no. of transmissions in t_{hd} for 10 CR cycles.

2.2. ProReact(Proactive and Reactive Handoff)

The ProReact segment is applied in the second phase, where it performs successful handoff to an idle channel. The SU performs proactive handoff based on a T.C.S. calculated by the Spectrum Broker through periodic sensing. The T.C.S. comprises of *n_pro* no. of channels, such that the total time incurred by individual SU in sensing all the channels in the T.C.S. using energy detection technique is less than $t_{handoff}$.

In the worst case, if all channels in T.C.S are sensed busy, the SU performs reactive handoff by randomly and uniformly selecting any available channel for sensing. The total no. of channels that can be sensed reactively is limited by the total handoff time allowed and is denoted by *n_react*.

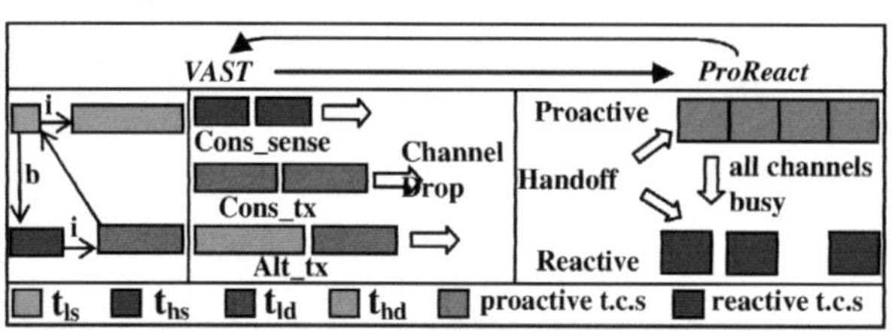

Figure 1. Schematic Diagram depicting the Proposed Algorithm

3. Mathematical Formulation

An analytical framework is designed to calculate the important call metrics for the proposed algorithm. Based on the exponential on-off traffic model for PU [2], the mathematical expression for channel idle probability during the Proactive phase of the *ProReact* algorithm is derived in (1), where β_k = transition rate from idle to busy period for *k*th channel.

$$Pr^{k}_{idle\_pro\ (t_2)} = e^{-(t_2)/(1/\beta_k)} \quad (1)$$

The probability of selecting the *k*th idle channel reactively from the list of available channels *l_react[]* at time t_2 is derived in (2) using the Renewal Theory. The average spectrum handoff delay at time t ($d^{(t)}_{h\_o}$) is derived in (3), where Pr_{h_o}, $P^{t_1}_{drop}$ = handoff and dropping probabilities at time t_1; $Pr^{k}_{idle\_pro(t2)}$, $Pr^{k}_{idle\_react\ (t2)}$ = probability of selecting idle channel k from *n_pro* and *n_react* respectively at time t_2; t_{s_i}, t_c, t_{sw}=channel sensing, consensus and switching time respectively.

$$Pr^{k}_{idle\_react\ (t_2)} = (1 - Pr^{k}_{busy\ (t_2)}) \sum_{\substack{i=1 \\ \forall A \in l\_react[]/k}}^{n\_react} \left[1/(A+1) \prod_{\forall v \in A} \left(1 - Pr^{v}_{busy\ (t_2)}\right) \prod_{\substack{\forall v' \notin A \\ \forall v' \notin i}} Pr^{v'}_{busy\ (t_2)} \right] \quad (2)$$

$$d^{(t)}_{h\_o} = \frac{1}{Pr_{h\_o}} \left[\sum_{k=2}^{n\_pro} \left\{ \left(k \times Pr^{k}_{idle\_pro(t_2)} \times Pr^{t_1}_{drop} \times t_{s\_i} \right) + t\_c + t_{sw} \right\} + \sum_{k=n\_pro+1}^{N} \left\{ \left\{ k \times Pr^{k}_{idle\_react(t_2)} \times Pr^{t_1}_{drop} \times n\_pro \times \left(t_{s\_i} + 1\right) \right\} + t\_c + t_{sw} \right\} \right]$$

$$\text{where} \quad N = n\_pro + n\_react \quad (3)$$

4. Simulation Results and Discussion

The proposed algorithm is implemented by designing a model for VoIP based CRN in MATLAB. The total no. of channels is 30. PUs follow exponential on-off model, and the on-off rates vary randomly. Total handoff limit for VoIP SUs is kept at 2s, and the CR cycle parameters are defined as, t_{ls}=50 ms, t_{hs}=100 ms, t_{ld}=100 ms, and t_{hd}=250 ms. Further, t_{s_i}=25 ms, t_c=100ms, and t_{sw}=25 ms. Based on previous works, three other cases are considered; i) React (Purely Reactive handoff [4]), ii) Pro-1 (Proactive Handoff to channel whose idle time is more than busy period [2]), and iii) Pro-2 (Variant of Pro-1:Proactive Handoff to channel whose idle time is 1.5 times more than busy period).

Two network scenarios are designed namely, i) best case: channel usage does not vary with time, and ii) worst case: channel usage varies drastically with time. It is observed from Figure 2(a) and (b) that the proposed algorithm selects maximum number of channels with higher idle probabilities, compared to other scenarios in the average and the worst case scenarios. Figure 2(c) shows maximum duration of successful transmission time for VoIP calls under the proposed algorithm, with a minimum 10% reduction in dropping probability, with respect to the other scenarios. Finally, as in Figure 3(a), *VAST* and *ProReact* segments ensure that handoff delay remains below 150 ms during proactive phase (40% lower than Pro-1 and Pro-2) and 600 ms in reactive phase (60% lower than React scenario).

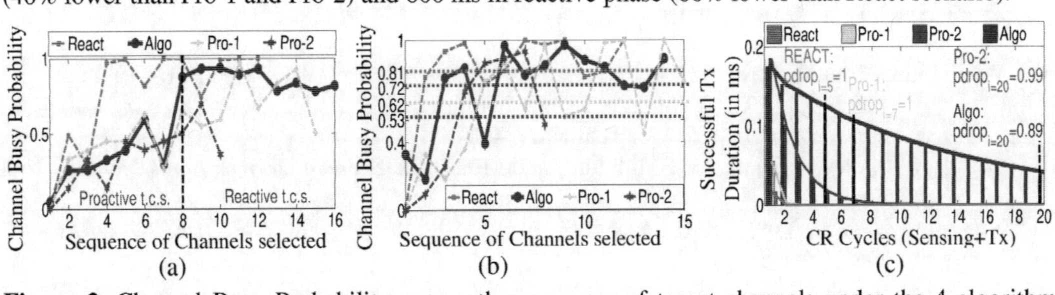

Figure 2. Channel Busy Probability versus the sequence of target channels under the 4 algorithms with respect to (a) best/average case, and (b) worst case scenarios; (c) Successful SU Transmission Duration with respect to the CR cycles under the 4 scenarios

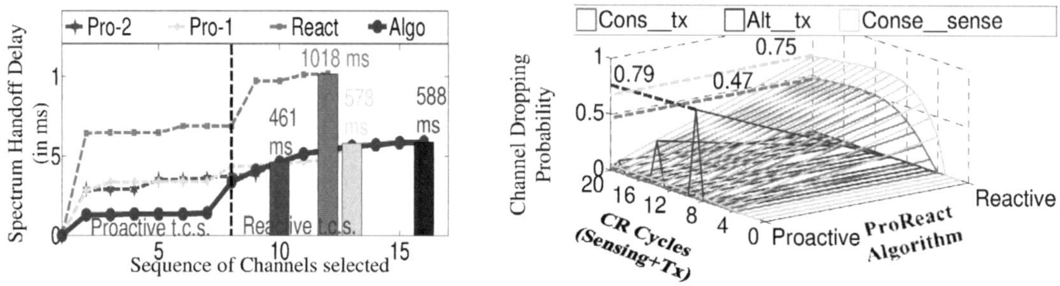

(a) (b)

Figure 3. (a) Average Spectrum Handoff Delay with respect to the different channels accessed under the 4 scenarios, (b) Variation in Channel Dropping Probability with CR cycles under Proactive and Reactive phases of ProReact

Figure 3(b) measures the no. of dropping instances after applying the proposed algorithm. Maximum no. of dropping instances is recorded due to *Alt_tx* condition during the Proactive phase of the *ProReact* algorithm, as the channel conditions are better in the Proactive phase (corresponding to the best/average case scenarios). On the other hand, both *Cons_sense* and *Cons_tx* conditions result in higher no. of dropping instances than *Alt_tx* in the Reactive phase of *ProReact* Algorithm.

5. Conclusion

Considering spectrum mobility as an integral part of CRN, this paper addresses the problem of spectrum handoff towards ensuring the QoS of VoIP users in the network. Accordingly, an efficient real-time spectrum handoff algorithm is proposed that not only guarantees successful handoff to the idle channels but also improves the QoS of existing VoIP calls. While the first phase of the algorithm (*VAST*) reduces call dropping probability by a minimum 10%, the second phase (*ProReact*) provides better channel selection mechanisms with higher idle probabilities. Overall, the proposed algorithm reduces the average handoff delay below the threshold limit of 200 ms, and maintains atleast 100 ms of effective communication duration for VoIP calls. Further studies are being carried out to design suitable prediction mechanisms for efficient channel selection.

Acknowledgement

The first author acknowledges the support of INSPIRE Fellowship 2012 from DST, Govt. of India.

References

[1] Akyildiz F, Lee W Y, Vuran M C and Mohanty S Sep. 2006 *Comput. Netw. J* (Elsevier) vol **50** no 13 p 2127–2159.

[2] Wang L, Wang C and Chang C Sep. 2012 *IEEE Trans. Mobile Comput.* vol **11** no 9 p 1499–1513.

[3] Zheng S, Yang X, Chen S and Lau C Dec. 2011 *IEEE Comm. Letters* vol **15** no 12 p 1312–1314.

[4] Wang C and Wang L Nov. 2012 *IEEE J. Sel. Areas Commun.* vol **30** no 10 p 2016–2028.

[5] Chuan P, Tran N H, Do C T, Moon S I and Choong H, Feb. 10–12. 2014 *Proc. Int. Conf. on Information Networking (ICOIN)* Thailand p 406–411.

[6] Davidson J, Peters J, Bhatia M, Kalidindi S and Mukherjee S *Voice Over IP Fundamentals*, 2nd ed. (Indianapolis, IN, USA: Cisco Press) chapter 1.

[7] Xiangwei Z, Ye L, Young K H and Soong A Nov.30–Dec. 4. 2008 *Proc. IEEE GLOBECOM* New Orleans LA USA p 1–5.

Cluster-Head Techniques for Single-hop Routing protocol in Energy Efficient Wireless Sensor Networks

Somasekhar Kandukuri, Nour Murad and Richard Lorion

Laboratory of Energy, Electronics and Process (LE^2P)
University de La Reunion 15 avenue Rene Cassin CS 92003
97744 Saint Denis CEDEX 9, Reunion - France

E-mail: somasekhar.kandukuri@univ-reunion.fr

Abstract. Energy efficient routing protocols are an active research area in Wireless Sensor Networks (WSNs). Like prior research works in efficient-efficient routing protocols, we enhance the LEACH cluster-head (CH) threshold mechanism. Our approach mainly focuses on the idea of maximizing the network lifetime on the basis of CH nodes remaining energy levels. In LEACH, CH changes occur during every round of the network irrespective of their energy levels. In this scenario, we propose an energy threshold function as well as an optimal CH threshold function to specify the dynamicity of the CH, whether to change the cluster-topology often or not with the consideration of its current energy levels. In this regard, simulation results show that the proposed mechanism can maximize the network lifetime effectively and delivers three times more data messages while comparing with the other known energy-efficient protocols from the literature.

1. Introduction

The advantages of Wireless Sensor Networks (WSNs) bring many potential applications, such as industrial control systems, vehicle tracking, home and office automation, water supplies, surveillance, wildfire monitoring, pollution monitoring, weather forecasting and enumeration are considered in the environmental monitoring applications [1–3].

In sensor networks, clustering protocols have the higher network performances in terms of network lifetime, less transmissions and low latencies than the traditional communication methods. In WSNs, there has been a great deal of research conducted on energy efficient clustering protocols, and several well-known methods have also been studied in the literature, such as [1, 4–11].

This paper presents a Single-hop Cluster-Based and Energy Efficient Protocol (SCEEP) Wireless Sensor Networks for prolonging the network lifetime based on Cluster-Heads remaining energy and optimal CH range levels. It is mainly considered to reduce the communication over-heads and unnecessary cluster-head changes at every round.

Section II describes the proposed SCEEP algorithm and features of the work are presented. Simulation results and comparisons are discussed in section III. Final section concludes the paper and suggestions for future work are given.

2. SCEEP Algorithm

In SCEEP, we consider a homogeneous sensor network model with an initial amount of energy for the sensor nodes. We used a first order energy model from [5] to design this work. Our approach mainly targets to reduce the communication overheads and unnecessary cluster-head (CH) changes at every round. In this proposal, we have been followed two kinds of clustering principle to elect the cluster-heads. The first principle follows LEACH algorithm of Eq.(1) to initialize the random sensor network at time interval t, and the mathematical expression of the CH probability threshold function is given as

$$P_i(t) = \begin{cases} \frac{K}{N-K(R \ mod \ N/K)} & C_i(t) = 1 \\ 0 & C_i(t) = 0 \end{cases} \tag{1}$$

Where K denotes the initial amount of optimal cluster head range, N is the total number of nodes in the network, R is the current round and, $C_i(t) = 1$ if the node i has not been already a cluster head in the last N/K rounds at given time instances t, otherwise $C_i(t) = 0$. According to Eq. (1), nodes always have the equal chances to become a cluster head for every round, regardless how much high or little amount of residual energy in the nodes have had. In this scenario, cluster topology or CH changes occur every round without considering the remaining energy levels of the nodes, which may lead to earlier nodes death and communication overheads or degrading the network lifetime. For avoiding the communication overheads and maximizing the network lifetime, we propose the second principle of clustering threshold function, which has been designed based on CHs remaining energy and fixed optimal cluster-head ranges at every round. However, the recent literature works have presented some advanced energy-efficient protocols in [9], [10] and [11], based on nodes residual energy of the network or using different sorts of nodes in the cluster-topology like a heterogeneous model, in which they still failed to reduce the communication and computational costs completely. In our approach, CHs check their energy status and optimal CHs range through the Algorithm 1 before calling for new cluster-heads request, as we can see it below. The proposed significant features of SCEEP protocol as follows

(i) Initial set-up phase of the network, we use LEACH cluster-head selection algorithm to elect the CHs during its first round.

(ii) After completion of the network first round (in seconds) simulations, current CHs check their energy levels $E_{CH_i(t)}$ to continue their intra-cluster communications with the nodes or not, by using energy threshold function E_{Th}. According to the network setup, the energy threshold value can vary based on the given initial energy levels of the nodes. If the current CH energy levels is greater than or equal to the energy threshold level, it continues its intra-cluster activities to receive the data among the cluster member nodes, otherwise CH destroys the current cluster and calls for the new cluster formation.

(iii) In this case, we fixed an optimal CHs threshold function $\hat{P}_{Th}$, if the CHs optimum level $\hat{P}$ is lower than the optimal threshold range, then it goes back to the Cluster-head election mechanism for electing new cluster-heads. In this regard, we fixed the optimal threshold range at 3.

(iv) If current CHs are higher than or equal to the minimum required CHs range, then it continues receiving the data from their corresponding member nodes.

Algorithm 1 Cluster-Head threshold functions

procedure : $Init(P_i(t), K, R, N)$
$E_{CH_i(t)}, E_{Th}, \hat{P}, and\ \hat{P}_{Th}$
Parameters: $t \leftarrow$ *node time instances*
$P_i(t) \leftarrow$ *CH Probability threshold function at time instance t*
$E_{CH_i(t)} \leftarrow$ *current CH remaining energy level*
$R(RoundNumber) \leftarrow$ *is the round iteration time in seconds*
$f(t) \leftarrow \sum_{i=1}^{N}\sum_{j=1}^{M} SN_i(X_j(t)),\ F(t) \leftarrow \sum_{k=1}^{L}\sum_{l=1}^{K} CH_k(X_l(t))$

    if $P_i(t) \leq K/(N - K(R \bmod N/K))$ then
        Start Cluster formation
        Count_CHs
    else if $E_{CH_i(t)} \geq E_{Th}$ then
        if $\hat{P} \geq \hat{P}_{Th}$ then
            Continue Data reception
        else
            Return to the network initialization $P_i(t)$
        end if
    else if $E_{CH_i(t)} < E_{Th}\ AND\ \hat{P} < \hat{P}_{Th}$ then
        Send `CH_candidates_clear()`
        Send `Clusters_Clear()` to destroy
        Return to the network initialization $P_i(t)$
    end if
end procedure

3. Simulation Results

In this simulation, SCEEP sensor network has been placed with 101 static sensor nodes, which are randomly distributed in a field of $400m \times 400m$. The initial energy of the node is $0.5\ J$ and each data transmission size is equal to $4\ kb$.

The results of SCEEP simulations are shown in figure 1(a) and (b). We presented some measurement metric analysis through both figures. According to the figure 1(a) and (b), the packet reception rate is 3 times more with SCEEP compared to DEEC and DDEEC. In figure 1(a),(b) and figure 2 SCEEP, where the First Node Dies (FND) metric shows at round 172 while other protocols FND are at round 28 and 36 respectively, based on the FND metric differences can clearly show that the early nodes depletion rate is 3 times lesser than the DEED and DDEEC protocols. When comparing with Half of the Nodes Alive (HNA) metric, we can clearly observe the differences, where the pointed line indicates in figure. 1(b) at 50 nodes alive, SCEEP extends the network lifetime 500 rounds ahead than the other protocols. From the below measurement analysis and proofs, SCEEP indeed have outperforming results than the other protocols. In fact, DEEC and DDEEC are developed with powerful nodes consideration of heterogeneous network model, such as normal and advanced nodes, even though SCEEP extends the network lifetime better than the other protocols.

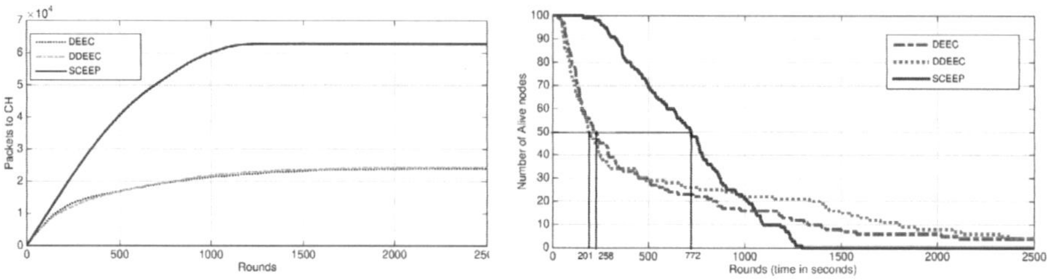

(a) (b)
Figure 1: (a) Number of data bits received and (b) Number of nodes alive over time.

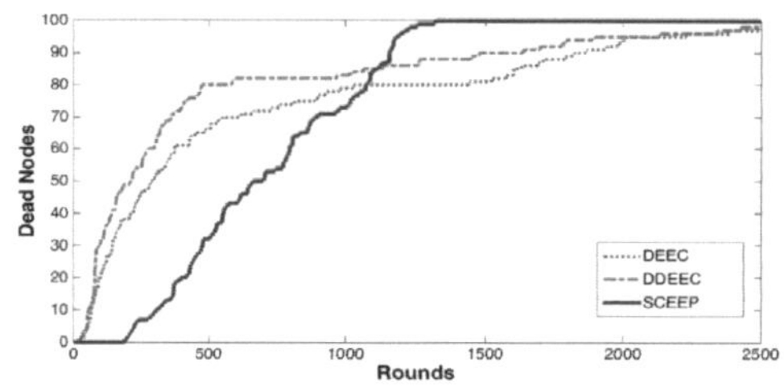

Figure 2: Number of depleted nodes over time.

4. Conclusion

A SCEEP distributed clustering protocol was presented, and the simulation results showed that the SCEEP have better performances than the DEEC and DDEEC protocols. In comparison, FND and HNA metrics demonstrate that SCEEP enhances the network lifetime and reduces depletion rate by a factor of 3 times. Then delivers 3 times more data than the other protocols. In addition, the further works on this paper that we are focused on to develop a new head node threshold function for initialization of the network or set-up phase clustering for multi-tier multi-hop heterogeneous sensor networks. Furthermore, we are also interested to deploy a sensor network using SCEEP on TelosB sensor nodes for measuring the system efficiency and throughput.

References

[1] Anastasi G Conti M Di Francesco M and Passarella A 7(3) pp. 537568
[2] Lotf J J Hosseinzadeh M and Alguliev R M 2010 IEEE in Computer Engineering and Technology (ICCET) 2nd International Conference on vol. 3 pp. V3650
[3] Enzinger M and Klein A 2012 in Sensor NodesOperation Network and Application (SN) vol. 25
[4] Younis O and Fahmy S 2004 Mobile Computing, IEEE Transactions on 3(4) pp. 366379
[5] Heinzelman W R Chandrakasan A and Balakrishnan H 2000 IEEE in System sciences Proceedings of the 33rd annual Hawaii international conference on pp. 10pp
[6] Lindsey S and Raghavendra C S 2002 IEEE in Aerospace conference proceedings vol. 3 pp. 31125
[7] Bandyopadhyay S and Coyle E J 2003 IEEE Computer and Communications in INFOCOM Twenty-Second Annual Joint Conference of the IEEE Societies vol. 3 pp. 17131723
[8] Siam M Z Krunz M and Younis O 2009 IEEE in INFOCOM pp. 621629
[9] Ibrahim R Ho Q-D and Le-Ngoc T 2013 IEEE 77th in Vehicular Technology Conference (VTC Spring) pp. 15
[10] Qing L Zhu Q and Wang M 29(12) pp. 22302237
[11] Elbhiri B Saadane R El Fkihi S and Aboutajdine D, 2010 IEEE in I/V Communications and Mobile Network (ISVC) 5th International Symposium on pp. 14

A Low-Cost RSSI-Based Localization System for Wildlife Tracking

Markus Hartmann, Thorsten Nowak, Lucila Patino-Studencki, Jörg Robert, Albert Heuberger and Jörn Thielecke

Friedrich-Alexander-Universität Erlangen-Nürnberg (FAU), Information Technologies, 91058 Erlangen, Germany

E-mail: markus.hartmann@fau.de

Abstract. This paper presents a low-cost localization system for the high-resolution tracking of bats. The system bases on a ground network consisting of multiple low-cost receiver stations, and ultra-lightweight transmitters mounted on the bats. A main challenge of the received signal strength based localization is the limited dynamic range of the employed low-cost receivers. This challenge is solved using an efficient 2-stage differential correlation concept. It significantly improves the dynamic wrt. conventional power detection methods. In addition, this concept requires low processing power and is robust wrt. frequency offsets. Finally, this paper presents a performance evaluation employing reference measurements recorded in the rain forest of Panama.

1. Introduction

Bats are important indicators for healthy natural environments. However, there is little knowledge of their behavior within their natural habitats, e.g. how they hunt or how they socially interact with other bats [1]. So far the detailed hunting behavior has only been analyzed in small areas using video cameras [2]. For monitoring significantly larger areas we have developed a new approach: The bats carry ultra-lightweight radio frequency transmitters ($< 2\,\mathrm{g}$) [3], and are then tracked by a ground network consisting of multiple receiver stations. For this purpose, the bat transmitters radiate special localization signals in the license-exempt 868 MHz band (915 MHz in the US). The ground network receiver stations, which consist only of cost-saving off-the-shelf components, are then able to localize the bat transmitters with a high temporal resolution of up to several Hz. However, the cost-saving components of this wireless sensor network (WSN) as well as the highly energy-constrained bat transmitters result in challenges. The most important challenges addressed in this paper are the limited dynamic range of the receiver and frequency offsets. Finally, the performance of our low-cost tracking system has been evaluated in a field trial in the rain forest of Panama. This paper is an extended version of the recent publication [4].

2. Wireless Sensor Network Architecture

Fig. 1 shows one possible realization of our bats tracking system. Four receiver stations are connected to a server via Ethernet cables. Each node is equipped with a directional antenna offering two patterns, a low-cost embedded processing platform (RaspberryPi B+) and two software-defined radio (SDR) capable terrestrial TV broadcast receivers. The SDR digitize the channel with a rate of 250 kS/s. Then, the digitized channel data of all receiver stations are streamed to a central server. This server uses a signal processing chain developed by ourselves that provides signal processing algorithms for the detection, the received signal strength (RSS) estimation, and finally the position estimation. This concept has the main benefit that all complex signal processing is located in a single server (e.g. a laptop computer), while the multiple receiver stations require low signal processing power.

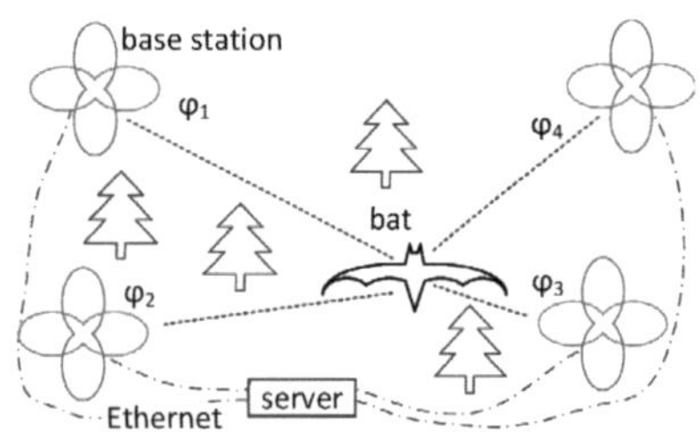

Figure 1. Bats localization system

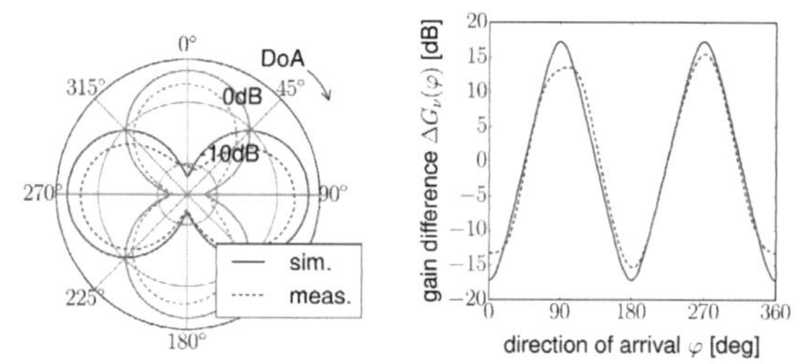

Figure 2. Simulated and measured antenna pattern and gain difference function

3. Antenna Array for RSSI-based Direction Finding

Direction finding in our system is based on the evaluation of the received signal strength indicator (RSSI) as presented in [5]. We are able to estimate the direction-of-arrival (DOA) using the RSS values of the two antenna patterns of a receiver station. The required directional antenna patterns result from a dipole array with a special feeding matrix. Details of the utilized antenna array and its optimization are discussed in [5]. Simulated and measured patterns of this antenna are presented in Fig. 2. The gain difference function, i.e. the RSS difference between the two antenna ports, $\Delta G_\nu(\varphi)$ is denoted by:

$$\Delta G_\nu(\varphi) \quad = \quad G_{RX1}(\varphi) - G_{RX2}(\varphi), \tag{1}$$

where G_{RX1} and G_{RX2} are the directional antenna gains, and φ is the DOA of the electromagnetic wave. At line-of-sight conditions with negligible multi-path the measured field strength difference is equivalent to the gain difference, i.e. $\Delta G_\nu(\varphi) \approx \Delta P_{RX}(\varphi)$. Thus, the DOA can be estimated evaluating ΔP_{RX}. As shown in Fig. 2 the measured gain difference function $\Delta G_\nu(\varphi)$ exceeds 14 dB for specific DOAs φ. This fact further increases the demands on the dynamic range: A valid estimation is only possible if the signals received via the two antenna patterns are not below the sensitivity nor exceeding the clipping level. We will discuss the impact on the localization area coverage in Sec. 5. Additionally, the RSS difference ΔP_{RX} is ambiguous with respect to the DOA as depicted in Fig. 2 (right subfigure). In our localization system we solve these ambiguities by sensor data fusion algorithms as presented in [6].

4. Signal Detection and Received Signal Strength Estimation

One main task of the central server is the detection of the bat signals within the multiple input streams digitized by the receiver stations. The dynamic range of the signal reception is limited by two bounds. The upper bound is the clipping in the ADC of the receivers. However, signals overdriving the ADC may still be decoded, but RSS measurements are not meaningful in that case. Evidently, the lower bound is the sensitivity of the receiver. Using the automatic gain control (AGC) of the SDR for extending the dynamic range is not possible, as the AGC is too slow for the short signals radiated by the bat transmitters.

Laboratory measurements indicated that simple signal detection algorithms, e.g. based on a sliding average power level detector and signal demodulation, do not lead to satisfactory results. The detection capability in combination with the 8-bit ADC leads to a dynamic range of less than 20 dB. This limited dynamic range results in a limited availability of the positioning as shown in Sec. 5. However, applying correlation-based detection overcomes this issue. According to the biologists the typical size of social groups of the Mouse-eared bat (*Myotis myotis*) is less than 12 individuals. Hence, the number of distinguishable signals is also limited by 12. Due to this reason we can correlate the received data against all possible 12 correlation sequences, resulting in excellent signal detection capabilities, even if the signal is below the noise level. For further limiting the processing load at the central server, we split this correlation into a 2-stage correlation as described in [4]. The first correlation stage detects the signals and reduces the amount of data. The second stage then provides precise RSS estimation and bat identification. Measurements show that we increase the dynamic range to 36 dB using our concept, i.e. 16 dB gain.

However, a common drawback of correlation-based signal detection algorithms is their sensitivity wrt. frequency offsets. The local oscillator (LO) of the bat transmitter has a frequency accuracy of 20 ppm, leading to a maximum frequency offset of 17.4 kHz at a carrier frequency of 868 MHz. Additionally, also the LO of the receiver shows such offset. The overall resulting frequency offset is too large for using standard correlators. In order to overcome this problem we use a correlation concept based on a differential correlator $L(\mu)$ [7]:

$$L(\mu) = \left| \sum_{k}^{L-1} r_{\mu+k} r^*_{\mu+k-1} s^*_k s_{k-1} \right|, \qquad (2)$$

where r_μ is the complex received signal, s_k is the transmitted sequence at sample k, L is the total sequence length, and $*$ denotes the complex conjugate operator. The differential correlator $L(\mu)$ translates frequency offsets into a constant phase rotation that has no impact on the further signal processing. In addition, the overall signal-processing load is reduced, as there is no need for complex frequency estimation in the receiver.

The RSS estimation error of the proposed 2-stage differential correlation is shown in Fig. 3. The solid line represents a laboratory measurement for frequency offsets of the receiver with respect to the carrier frequency of the transmitter. The measurement shows a maximum RSS for an offset of approximately 25 kHz, which represents the total frequency offset between receiver LO and transmitter LO in this particular measurement. The dashed and dotted lines show simulation results for different frequency offsets without and with an initial frequency offset of 25 kHz. Up to a frequency offset of 40 kHz the RSS estimation error is < 1 dB, which is sufficient for our targeted application.

5. Area Coverage and Localization Results

The coverage in the presented scenario is defined by the number of receiver stations that are able to measure a valid RSS difference ΔP_{RX}. As previously mentioned, if one of the two broadcast receivers of a receiver station is clipping or receiving below its sensitivity level, no valid RSS difference can be estimated. Fig. 4 and 5 indicate how many receiver stations are able to receive a valid RSS difference in an area of 100×100 m. If we assume a dynamic range of 20 dB we only have a small part of the area where all four receiver stations are able to provide valid DOA measurements. We also have regions where only one base station is able to deliver a valid DOA, resulting in a low accuracy and potential position ambiguities. Using our 2-stage correlator concept, we are able to increase the dynamic range to 36 dB. This results in a significantly increased coverage as depicted in Fig. 5. At all positions at least three valid DOA are available. Thus, we get good localization accuracy and do not suffer position ambiguities. Fig. 6 shows

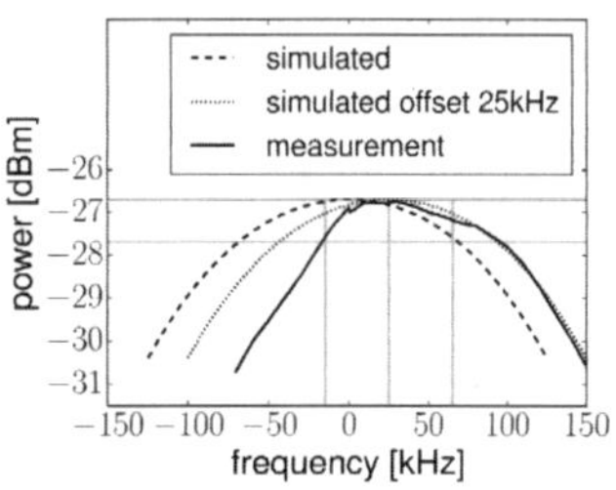

Figure 3. received signal strength estimation error

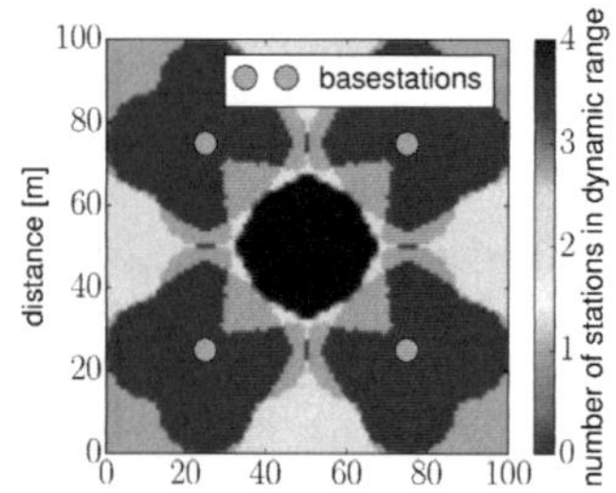

Figure 4. Coverage map at a dynamic range of 20 dB

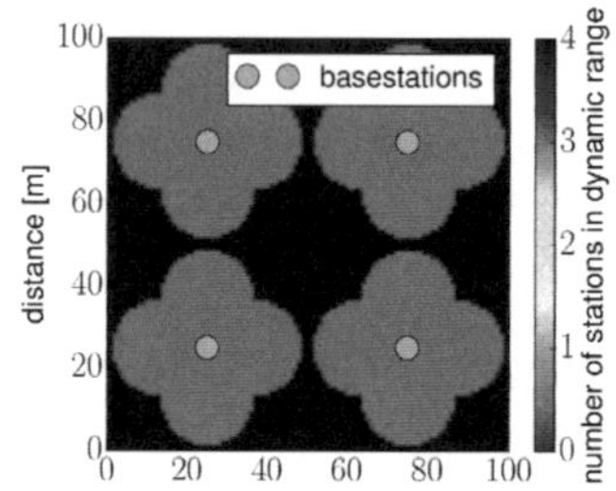

Figure 5. Coverage map at a dynamic range of 36 dB

a measurement in the rain forest of Panama with heavy multi-path effects. A bat transmitter carried by hand has been used during these measurements. The position estimation has been performed utilizing three receiver stations. The resulting average error is approx. 3.4 m.

6. Conclusion

In this paper we presented a low-cost localization system for the tracking of bats. We proved that our system applying RSS-based DOA estimation is able to cover an area of 100 × 100 m. The major challenges of our low-cost system, i.e. dynamic range and frequency offsets, have been solved by means of our 2-stage correlation concept. Results of our Panama field trial underline the performance and the precision of our system.

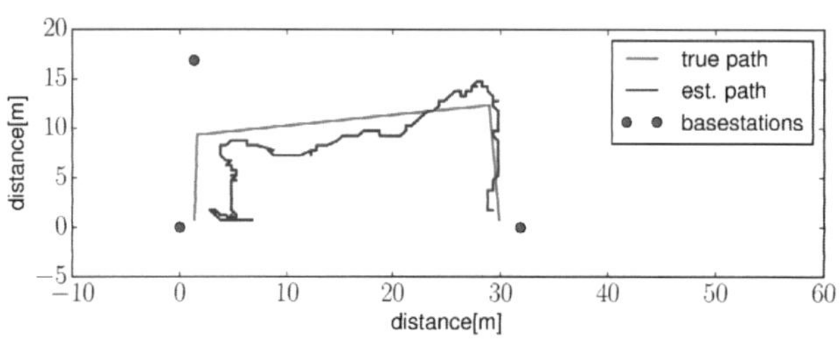

Figure 6. Estimated and true trajectory for a reference measuement in a rain forest area

Acknowledgments

This work is supported by German Research Foundation DFG grant FOR 1508, Research Unit BATS.

References

[1] Ripperger S, Tschapka M, Kalko E, Rodrıguez-Herrera B and Mayer F 2014 *Agriculture, Ecosystems and Environment* **185** 9–15
[2] Holderied M W, Baker C J, Vespe M and Jones G 2008 *Integrative and Comparative Biology* **48** 74–84
[3] Hierold M, Ripperger S, Josic D, Mayer F, Weigel R and Koelpin A 2015 Low-weight wireless sensor network for encounter detection of bats *IEEE Topical Conference on Wireless Sensors and Sensor Networks*
[4] Hartmann M, Nowak T, Robert J, Tröger H M, Heuberger A and Thielecke J 2015 A low-cost rssi based localization system design for wildlife tracking *Radio and Antenna Days of the Indian Ocean 2015*
[5] Hartmann M, Pfadenhauer O, Patino-Studencka L, Tröger H M, Heuberger A and Thielecke J 2015 Antenna pattern optimization for a rssi-based direction-of-arrival localization system *ION's Pacific PNT Conference 2015*
[6] Hartmann M, Nowak T, Pfadenhauer O, Thielecke J and Heuberger A 2016 A grid-based filter for tracking bats applying field strength measurements *Proceedings of the 2016 IEEE/IFIP Wireless On-demand Network systems and Services Conference* pp 184–191
[7] Choi Z Y and Lee Y H 2002 *Communications, IEEE Transactions on* **50** 1062–1065 ISSN 0090-6778

IOP Conf. Series: Materials Science and Engineering **120** (2016) 012005 doi:10.1088/1757-899X/120/1/012005

Low-cost inkjet antennas for RFID applications

T Çiftçi, B Karaosmanoğlu, and Ö Ergül

Department of Electrical and Electronics Engineering, Middle East Technical University, Ankara, Turkey

E-mail: {tciftci,bariscan}@metu.edu.tr, ozgur.ergul@eee.metu.edu.tr

Abstract. We present paper-based inkjet antennas that are fabricated by using silver-based cartridges in standard printers. In addition to their low costs, the produced antennas are flexible, environmentally friendly, and suitable for radio-frequency identification (RFID) applications. Among alternative choices, hybrid structures involving loop and parasitic meander parts are preferred and successfully combined with passive RFID chips. We also discuss main challenges in the design and fabrication of low-cost inkjet antennas and the related RFID tags.

1. Introduction

Inkjet antennas that are produced by printing metals on papers and similar substrates have attracted great interest due to their favorable characteristics, especially those making them suitable for radio-frequency-identification (RFID) applications [1]–[6]. While inkjet antennas can be produced by using special material printers [3], their fabrication costs can be further decreased by using silver-based cartridges in standard printers. Hence, using a simple setup, one can fabricate inkjet antennas that are relatively inexpensive, flexible, and environmentally friendly. On the other hand, many parameters affect the quality of such inkjet antennas, especially when they are fabricated by using low-cost standard printers. Therefore, these parameters need to be considered carefully and optimized for producing high-quality RFID tags.

In this work, we present low-cost inkjet antennas that can be used in passive tags for RFID applications. The developed antennas involve loop parts that are directly connected to the chips and parasitic meander parts to increase the gain, hence the range of the tags. We also discuss the challenges in in-house fabrications of such inkjet antennas and their usages in RFID tags. These include printing parameters, post-processing, connection problems, and stability issues in the overall production cycles. Finally, we present a type of low-cost RFID tags produced by using fabricated antennas.

2. Low-Cost Inkjet Printing

The setup that we use for modeling, fabricating, and testing low-cost inkjet antennas is depicted in Fig. 1. After an antenna is designed and analyzed in a rigorous simulation environment based on integral-equation formulations, a black copy of the antenna geometry (antenna surfaces) is generated for printing. We use standard Epson printers and silver-based cartridges for printing antennas. A preferred ratio of the silver in the cartridge content is 25%, while both lower and higher ratios may lead to lower qualities, depending on the paper type and curing. Photograph papers that are resistant to high temperatures are used as substrates. After printing, each

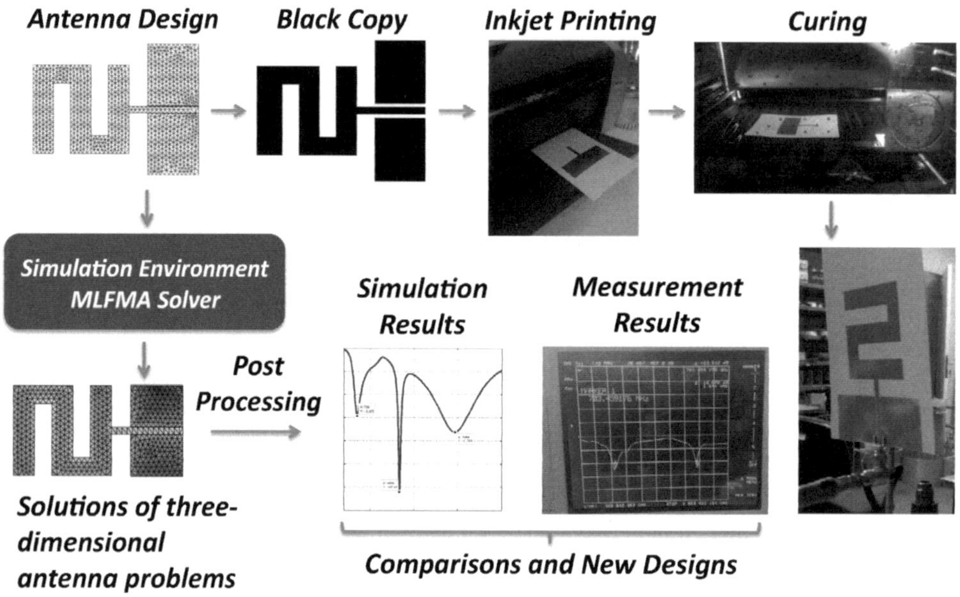

Figure 1. A summary of the setup for modeling, fabricating, and testing low-cost inkjet antennas.

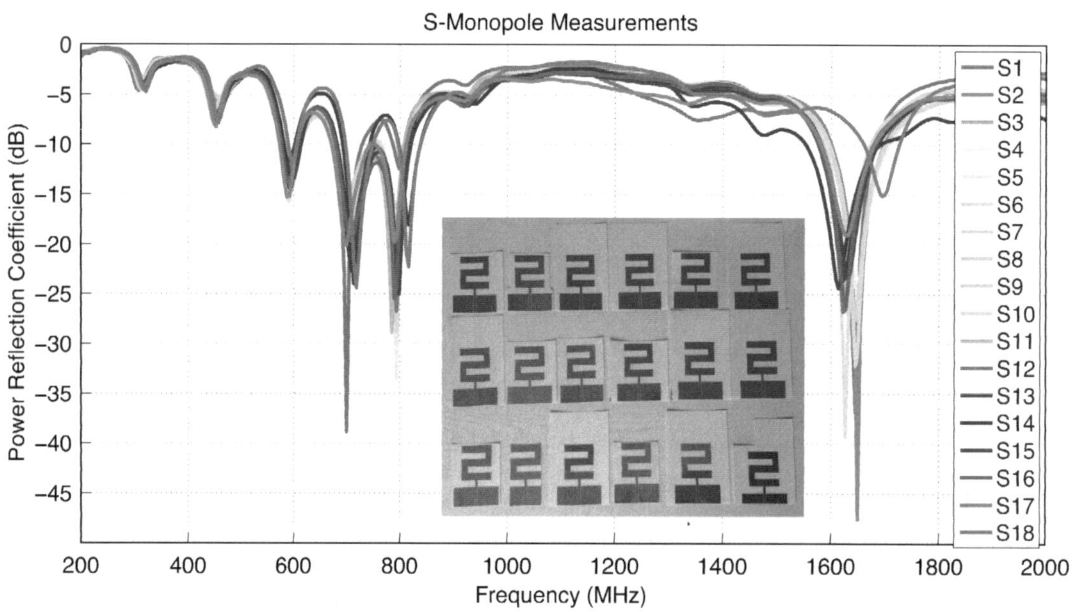

Figure 2. Measured power reflection (when matching to 50 Ω) values for S-monopole antenna samples that are produced consecutively.

antenna is cured in a simple oven at 150°C for at least 1 hour. The printer heads need to be cleaned after each printing session to avoid residues that can easily oxidize. On the other hand, the quality of the prints easily deteriorates and the printer (or the printer head, if possible) need to be replaced regularly.

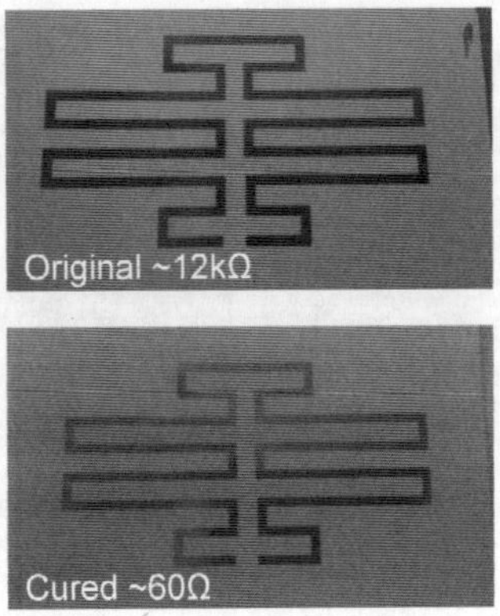

(a)

(b)

Figure 3. (a) Effect of temperature curing. (b) A loop/meander antenna designed and fabricated for passive tags in RFID applications. The loop part has terminals that can directly be connected to RFID chips. The parasitic meander part is required for increasing the readability distance.

Despite all efforts to standardize inkjet printing, the antennas produced simultaneously often exhibit different electrical characteristics, especially different input impedances with respect to frequency. Fig. 2 presents measurement results for 18 different S-monopole antennas that are produced consecutively. These inconsistencies are mostly caused by unavoidable built-in parameters of the printers that are optimized for printing standard inks rather than silver-based material. Connecting the produced antennas to coaxial cables for measurements is another important issue, which is also a reason for the inconsistencies in Fig. 2. Soldering and physical pressures easily damage the antenna; hence, using conducting epoxy for electrical connections is essential for sensitive measurements. The epoxy is also used for connecting RFID chips to antennas. This needs a second curing session at a suitable temperature. Fig. 3(a) depicts the effect of an optimal temperature curing on a long conductive line, while Fig. 3(b) demonstrates the flexibility of the resulting prints.

3. Measurement Results

Fig. 4 presents measurement results for a low-cost inkjet antenna involving loop/meander parts. In the same figure, a photograph of the antenna with a 1 Turkish Lira coin is also included. As shown in the photograph, an RFID chip can directly be connected to the antenna terminals (loop part) without a matching circuit. The input impedance of the used chip is around $20.55-191.2i$ Ω (where $i = \sqrt{-1}$) at 865–868 MHz, which is a given frequency range for commercial RFID readers. Hence, the input impedance of the antenna needs to be around $20.55 + 191.2i$ Ω in the same range. Fig. 4 depicts the measured input impedance of the antenna with respect to frequency from 50 MHz to 6 GHz. In addition, the power reflection coefficient values (in dB) when the antenna is matched to $20.55 - 191.2i$ Ω are plotted. The power reflection graph when the antenna is matched to the chip (not shown) is slightly different, as the capacitive part of the chip impedance depends on the frequency. However, we prefer the reflection when the antenna needs to be matched to a fixed impedance ($20.55 - 191.2i$ Ω in this case) for re-scaling purposes. In a zoomed plot, it can be seen that the antenna perfectly matches to the chip with less than -18 dB power reflection in the required frequency range. We emphasize that the plots in Fig. 4 are based on an average behavior of the designed antenna over 8 different samples and their measurements since the exact characteristics of each printed antenna tend to be different.

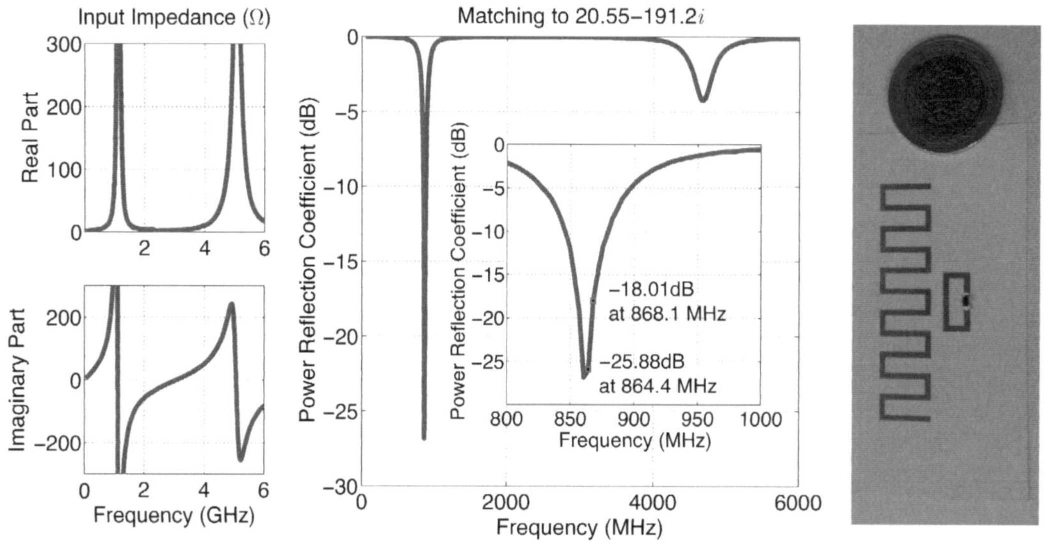

Figure 4. Measured input impedance and power reflection (when matching to $20.55 - 191.2i$ Ω) values for a loop/meander antenna. The antenna when connected to an RFID chip is also shown. The meander part has little effect in the input impedance, but its distance to the loop greatly changes the range of the RFID tag. This distance (around 0.01–2 m) is easy to optimize by printing loop and meander parts separately and performing a reading test.

4. Concluding Remarks

We present low-cost inkjet antennas that are produced by using silver-based cartridges in standard inkjet printers. A loop/meander antenna, which is compatible with commercial RFID chips, is particularly focused. The cost of each RFID tag involving the proposed antenna and a chip is less than 1 USD, including *all processes* from design to the end product in the laboratory environment. While this price is already acceptable for many industrial applications, it can be further reduced if the antennas are produced in large amounts. While the developed fabrication cycle enables an easy production of inexpensive, flexible, and green antennas, it needs to be optimized for high-quality tags. The developed antennas also need design margins as the standard printers may lead to silver prints with identical geometries but different electrical characteristics. We show that, by using a simple loop/meander combination, it is possible to achieve −18 dB average matching to a capacitive chip.

Acknowledgments

This work was supported by the Scientific and Technical Research Council of Turkey (TUBITAK) under the Research Grant 113E129 and via the 1512 Support Program (project no. 2130002), and by a BAGEP Grant from Bilim Akademisi (The Science Academy, Turkey).

References

[1] Nikitin P V, Lam S and Rao K V S 2005 *Proc. IEEE Antennas and Propagation Soc. Int. Symp.* 353
[2] Yang L, Zhang R W, Staiculescu D, Wong C P and Tentzeris M M 2009 *IEEE Antennas Wireless Propag. Lett.* **8** 653
[3] Rida A, Yang L, Vyas R and Tentzeris M M 2009 *IEEE Antennas Propag. Mag.* **51** 13
[4] Cook B S and Shamim A 2012 *IEEE Trans. Antennas Propag.* **9** 4148
[5] Subbaraman H, Pham D T, Xu X, Chen M Y, Hosseini A, Lu X and Chen R T 2013 *IEEE Antennas Wireless Propag. Lett.* **12** 170
[6] Maza A R, Cook B, Jabbour G and Shamim A 2012 Microwaves, Antennas & Propagation, IET **6** 1366

IOP Conf. Series: Materials Science and Engineering **120** (2016) 012006 doi:10.1088/1757-899X/120/1/012006

RF Pulse Signal Integrity Analysis for Nonlinear Ended Microstrip Line Atom-Probe Tomography

L. Zhao[1], A. Delamare[1], A. Normand[1], F. Delaroche[1], O. Latry[1], F. Vurpillot[1], B. Ravelo[2]

[1] UMR CNRS 6634 - Normandy University GPM, Université and INSA de Rouen, Saint-Etienne-du-Rouvray, France

[2] IRSEEM EA 4353, at the Graduate School of Engineering, ESIGELEC, Saint-Etienne-du-Rouvray, France

E-mail: lu.zhao@univ-rouen.fr

Abstract. A signal integrity (SI) analysis of high voltage rectangular short pulses for the atom-probe system is explored in this paper. The operated RF transient pulse is considered for exciting on material sample inside an ultra-high vacuum (UHV) cryogenic chamber. The ns-duration pulse signal is injected into the cryogenic analysis chamber through the transmitting system mainly constituted by a microstrip interconnect line ended by optical controlled nonlinear load. The whole system frequency characterization is performed based on the S-parameter measurements. As expected, a challenging ultra-short rectangular shape pulse is exhibited by the pulser. Promising experimental results with the improvement of ion mass spectrum is demonstrated with the designed RF pulser.

1. INTRODUCTION

Atom probe tomography is a nanoscale analysis technique for material science. It is now widely spread in the nano-characterization world. One of the key performances of the modern instrument is its mass resolving power. Indeed, recent study on atom-probe [1] demonstrated that a ns-duration rectangular voltage pulse could improve the analysis capability of this scientific instrument. Different techniques can be deployed to generate such a short duration pulse. A pulse generator concept [2] using photoconductive switches has been generally used for producing ultra-short pulses with rise-time in ps-scale and sub-ns duration [3], [4]. In the field of pulsed power technique, the rise-/fall time is mainly depended on the switching speed of high power switch. The optical technique has more advantage than conventional electronic techniques. Photoconductive semiconductor switches thus are very suitable candidate for the high-power pulsed application because they are simple, faster switching speed and jitter-free response etc. Among the numbers of PCSS material, the silicon PCSS is chosen for the application because of its linear response and simple integration[5].

In this paper, an improved atom-probe system constituted by a rectangular shape pulse generator presenting ns full width is examined. The signal integrity (SI) investigation will be carried out by considering the S-parameters and the transient behaviours of the transmitted ns-duration FW pulse.

2. CONFIGURATION OF THE ns-PULSE GENERATOR ATOM-PROBE UNDER STUDY

A needle-like sample introduced in a vacuum chamber (10^{-10}mbar) and maintained low cryogenic temperatures (~80K), the atoms on the sample surface are eroded by means of negative HV pulses applying on the local electrode which is placed very close to the sample (<1mm) [1][6]. A position-sensitive detector in front of the sample (10-20cm) permits identifying the atomic specie by the time-of-flight mass spectrometry [6]. Due to the stochastic nature of the field evaporation process (the physical phenomenon inducing the atom by atom erosion), it degrades the mass resolution of atom-probe. Therefore, the time-domain behaviour of the field evaporation which can be controlled with the excitation pulse voltage can be a key element for the atom-probe improvement. The innovative pulse generator solution with challenging rise-/fall-time investigated in this paper is mainly based on the nonlinear loaded microstrip transmission line (TL) using photoconductive switch (PCSS). An HV rectifier diode sustaining up to 4kV DC is employed and associated with an amplified Ti: Sa fs laser (provided by Quantronic IntegraE®) served as an optical command. The ns-pulse generator under study is overviewed in Figure 1. The PCSS ended microstrip TL printed on the FR4 substrate having relative permittivity ε_r=4.4 and thickness h=1.6mm can be seen in left of Figure 1. The V_0=6kV DC driven source resistance was fixed to R_s=10kΩ. In difference with the generator proposed in [7], a nonlinear load constituted by the PCSS is inserted in upstream of the termination R_0=50Ω. The PCSS is controlled by an optical beam in order to deliver the stored energy into the load resistance. Then, a ps range rise time electric pulse can be generated across R_0 [7].

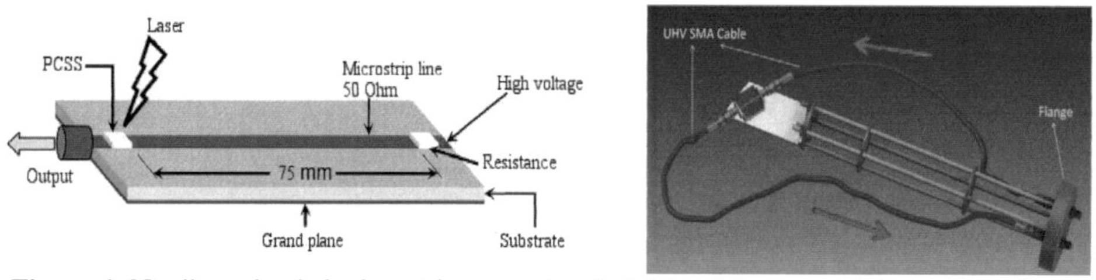

Figure 1. Nonlinear loaded microstrip generator (in left) and the transmitting system (in right).

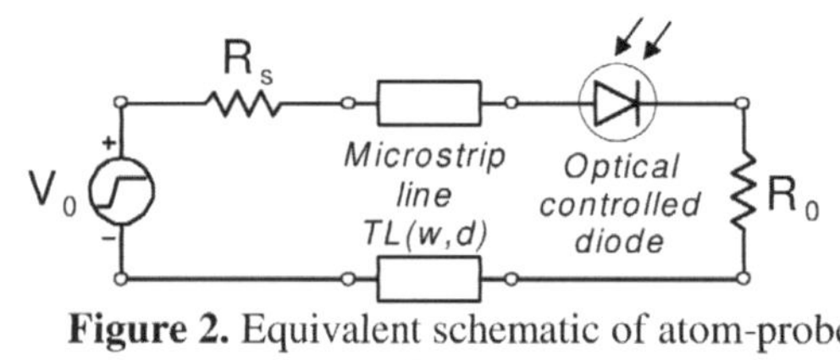

Figure 2. Equivalent schematic of atom-probe pulser.

The ns rectangular pulses can be obtained with laser energy 20μJ/pulse. The pulse generator output is connected to the input of the transmitting system as depicted in the right of Figure 1. For the basic SI analysis, the pulse generator microstrip structure was assumed as equivalent to the electrical circuit schematized in Figure 2.

As illustrated in right of Figure 1, the proposed atom-probe generator transmitting pulse system is essentially composed of the microstrip circuit interconnected to the UHV SMA cables. The microstrip TL is laminated with Copper conductor presenting 35μm thickness having physical width w=3mm and length d=75mm. The centre of the microstrip TL is pierced with via presenting diameter Ø=300μm. This hole serves as a microelectrode, and the sample directly be excited by the electrical pulse, atoms are fielded evaporated from the sample and are projected through the hole to the position sensitive detector placed about 20cm in front of the device.

3. INVESTIGATION ON THE OPERATED ns-DURATION PULSE SIGNAL INTEGRITY WITH S-PARAMETERS ANALYSIS

With the previously described generator, we expected to exhibit 1ns full width half maximum (FWHM) rectangular shape voltage to excite the material sample. To do this, the PCSS was excited with fs duration laser pulse. The different blocks of the generator S-parameter responses and the time-domain plot of the pulse will be explored in the next paragraphs.

3.1. ns-Duration Rectangular Shape Pulse Generation

First and foremost, after injecting the slightly distorted rectangular shape pulse plotted in left of Figure 3-1 into the UHV chamber by the transmitting system, the output signal plotted in dashed line was observed. It is noteworthy that the measurement was performed with the Agilent® infiniium 54855A oscilloscope. As the transmitting system behaves as low-pass filter behavior, the high order harmonics was attenuated. The corresponding spectra obtained from Fast Fourier Transform (FFT) of measured signal from oscilloscope are displayed in the right of Figure 3-1. It illustrates that the system loss occurs from 1.5GHz, and due to the oscilloscope bandwidth limit, the spectra is vanished from 6GHz.

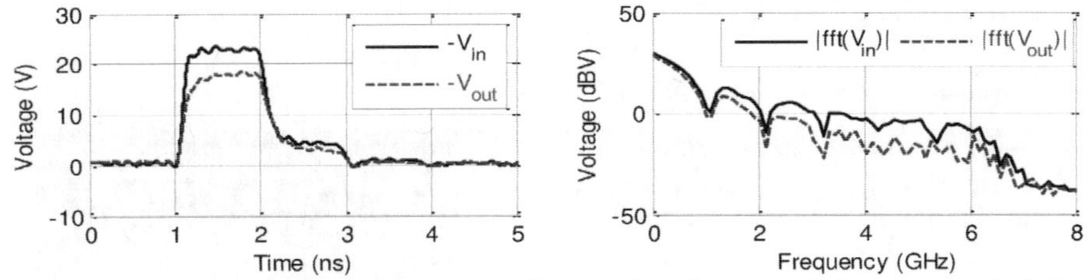

Figure 3-1. Transient signals before and after passing the transmitting system: dash line is the input and dash dot line is the output and the corresponding FFT.

3.2. Transmitting System Characterization

To investigate the frequency limitation and matching problem of assembled two-port transmitting system, the return and insertion losses S11 and S21 were measured up to 8GHz by using the ENA E5071B VNA for the UHV SMA cables, the flange and the 2-port transmitting system, respectively. The frequency dependent results are displayed in the Figure 3-2. Note that the insertion loss of the whole system is due to limitation of the flange and the vacuum cables at high frequency range. Moreover, S_{21} is below -3dB above 1.5GHz. It means that most of energy (more than 50%) was not transmitted over the upper frequency band.

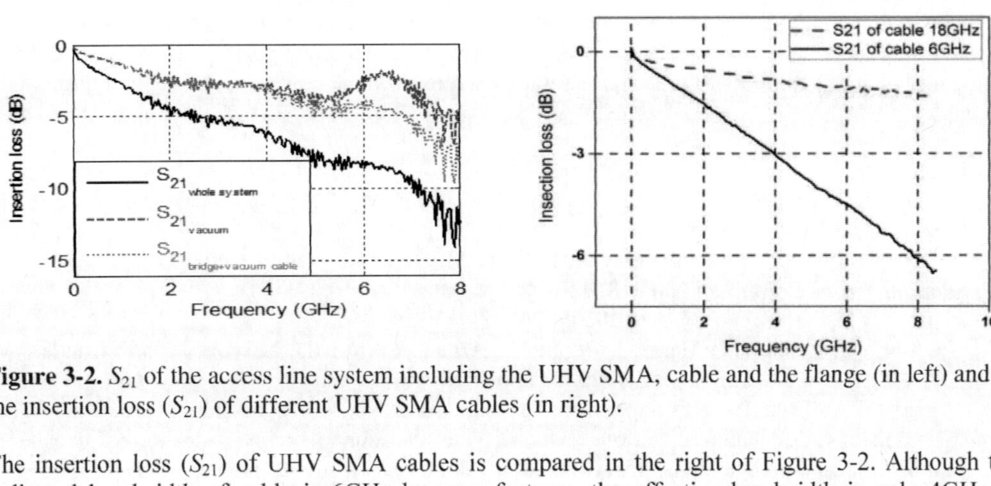

Figure 3-2. S_{21} of the access line system including the UHV SMA, cable and the flange (in left) and the insertion loss (S_{21}) of different UHV SMA cables (in right).

The insertion loss (S_{21}) of UHV SMA cables is compared in the right of Figure 3-2. Although the indicated bandwidth of cable is 6GHz by manufacturer, the effective bandwidth is only 4GHz in practice.

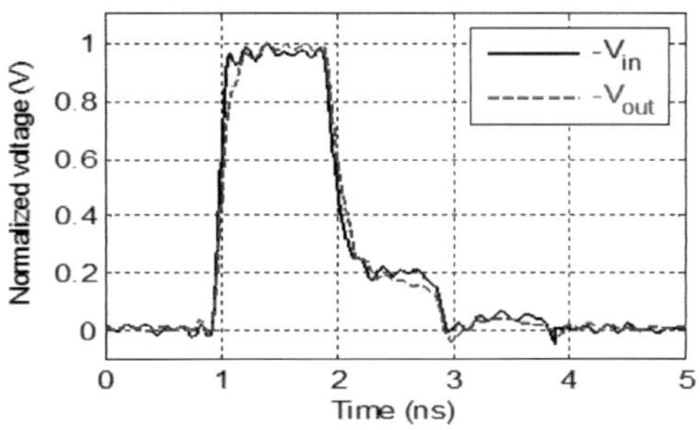

Figure 3-3. Signals measured after replacing the UHV SMA cable.

Considering all the system, we proposed to replace the UHV SMA cables with another one having much larger bandwidth of 18GHz as a simple way to improve the transmission shown in the right of Figure 3-2. As expected, a significant improvement was realized which can be seen with the transient plot shown in Figure 3-3. Practical transmission efficiency with rectangular shape pulse presenting rise time better than 200ps is realized.

3.3. Mass Spectrum Improvement Application

Although rectangular pulses with ns duration are not very well transmitted by the system, the voltage pulse applied to the sample could be a Gaussian-similar pulse. Preliminary mass spectrum obtained by analysing Al sample (M=27uma) with standard voltage pulse and with sub-ns short pulses demonstrates a significant improvement of the mass resolution power. Comparing to the standard atom-probe, the pulse transmitting system installed in atom probe displayed in Figure 4. The experimental result shows that the mass resolution is 3 times better than that obtained by the standard atom probe.

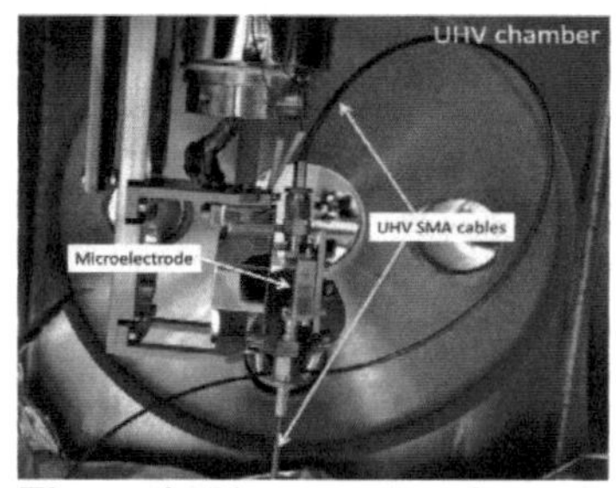

Figure 4. Photography of atom probe analysis chamber with proposed microstrip TL (microelectrode).

4. CONCLUSION

A new method to improve the atom-probe tomography based on RF transient ns-pulse generator using nonlinear ended microstrip TL with PCSS is developed. The transmitting system has been investigated in frequency domain. The signal distortion is largely due to the combined effect of the UHV SMA cable and the flange. To avoid the transmission problem, a sub-ns rectangular shape pulse generator adapted to the UHV chamber is needed so that the generated voltage pulses can be applied directly to the sample.

5. REFERENCE

[1] Zhao L, Normand A, Delaroche F, Ravelo B and Vurpillot F 2015 *Int J Mass Spectrom* **386** pp 47–53
[2] Vergne B, Couderc V, Barthélémy A, Lalande M, Bertrand V and Gontier D 2007 *Dig Tech Pap Int Pulsed Power Conf* pp 818–823
[3] Auston D H 1975 *Appl Phys Lett* **101** pp 24–27
[4] El Amari S, Kenaan M, Merla C, Vergne B, Arnaud-Cormos D, Leveque P and Couderc V 2010 *IEEE Photonics Technol Lett* **22** (21) pp 1577–1579
[5] Amari S E L 2010 Développement et caractérisation de générateur optoélectronique d'impulsions de champ électrique nanoseconde et subnanoseconde de forte intensité : application au domaine biomédical (in French), *PhD thesis*, Univ. Limoges, France
[6] Miller M K 2014 *Atom Probe Tomography: the Local Electrode Atom Probe* **10** (S02)
[7] Lee C H 1990 *IEEE Trans Microw Theory Tech* **38** (5) pp 596–607

IOP Conf. Series: Materials Science and Engineering **120** (2016) 012007 doi:10.1088/1757-899X/120/1/012007

Compact Wide Stopband Lowpass Filter with High Suppression Using Corrugated Transmission Line

P M Raphika, P Abdulla and P M Jasmine
Division of Electronics, School of Engineering, Cochin University of Science and Technology, Kochi-22, Kerala, India

E-mail: pmraphika@yahoo.co.in

Abstract. Compact and wide stopband lowpass filter with high suppression level using corrugated high impedance transmission line with symmetrically loaded multiple patch resonators is designed and demonstrated. The measured results validate the simulated characteristics and exhibit a roll-off of 60 dB/GHz and a wide stopband with suppression level better than 31 dB from 2.65 GHz to 13.4 GHz. The 3 dB cut-off frequency of the filter is at 2.04 GHz with passband insertion loss of 0.7 dB. The normalized circuit size of the developed filter is 0.21824 λ_g x 0.1314 λ_g, where λ_g is the guided wavelength of 80.646 mm at cutoff frequency for 50 Ω impedance.

1. Introduction

Compact lowpass filters with wide stopband bandwidth and high suppression level are in great demand for modern communication systems to suppress harmonics and spurious signals. Planar filters exhibit good electrical and mechanical characteristics together with the ease of fabrication, which makes it suitable for radio frequency and wireless applications [1]. Various methods have been proposed to exploit a lowpass filter with sharp roll-off and wide stopband by utilizing different dielectric material properties. The technique of etching defects in the ground plane supports sharp switching characteristics by suffering additional radiation due to the defected ground [2]. V. K. Velidi *et al.* reported the method of stub-loaded coupled-line hairpin unit to develop sharp roll-off lowpass filter with wide stopband [3]. Though size reduction was achieved by the proposed design, applications of the filter are limited in the S-band. Design of sharp roll-off lowpass filter with ultra wide stopband by creating multiple transmission nulls was proposed in [4]. M. Hayathi *et al.* developed a compact lowpass filter with wide stopband using modified semi-elliptic and semi-circular microstrip patch resonators [5]. To develop lowpass filters with good frequency characteristics the method of loading multiple patch resonators on high impedance transmission line has been reported recently by the authors in [6–7]. Although these reported filters are good choice for modern communication systems, they suffer deficiencies such as limited stopband suppression level and large circuit size.

In this paper, a compact microstrip lowpass filter with wide stopband and high suppression level is designed by loading resonator1 as centre resonator, and resonator2 and resonator3 as side resonators on corrugated high impedance transmission line. Each resonator is designed with high impedance short circuited stub and low impedance open circuited patch. The 3 dB cut-off frequency of the filter is at 2.04 GHz with a roll-off of 60 dB/GHz. The filter achieves a high stopband suppression level better than 31 dB from 2.65 GHz to 13.4 GHz. The wide stopband characteristics with high suppression level has simultaneously obtained by applying a narrow band design for the resonator1 and wideband design for resonator2 and resonator3.

2. Proposed filter design

Sharp switching characteristics of a lowpass filter from passband to stopband can be realized by generating infinite attenuation poles at finite frequencies. By cascading multiple resonators on a high impedance transmission line we can execute such characteristics [1]. As shown in figure 1, the proposed filter consists of a centre resonator, resonator1 and two side resonators, resonator2 and resonator3 respectively. Resonator1 is designed with high impedance short circuited stub and modified funnel patches. U-shaped high impedance stubs and modified triangular patches are used for the design of resonator2. These patch resonators are loaded as $\lambda g/2$ stepped impedance resonators on a high impedance corrugated main transmission line with characteristic impedance of 120 Ω.

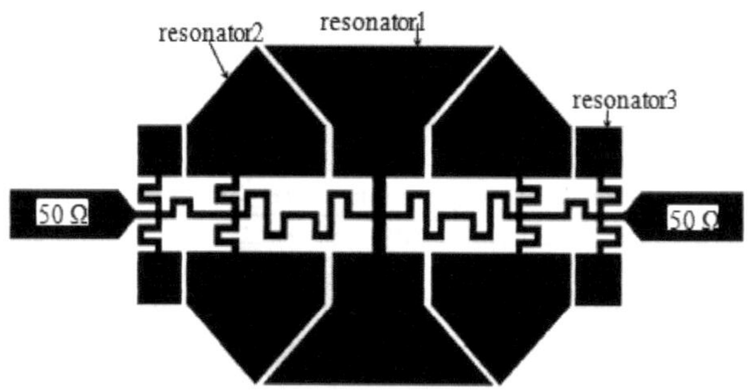

Figure 1. Schematic structure of the proposed filter.

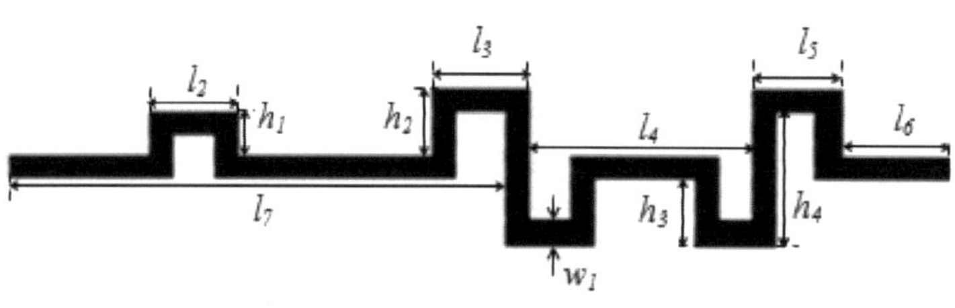

Figure 2. Layout of the corrugated high impedance transmission line.

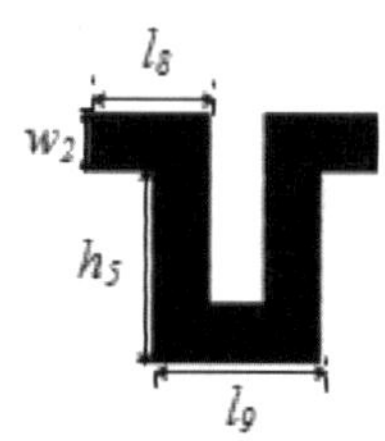

Figure 3. Layout of U-shaped high impedance stub of resonator2.

Figure 2 demonstrates the layout of the corrugated transmission line symmetrical about longitudinal axis. The optimum dimensions selected for the transmission line are: l_1 = 0.75 mm, l_2 = 0.8 mm, l_3 = 0.9 mm, l_4 = 2.1 mm, l_5 = 0.8 mm, l_6 = 1 mm, w_1 = 0.2 mm, h_1 = 0.4 mm, h_2 = 0.65 mm, h_3 = 0.65 mm and h_4 = 1.3 mm. As illustrated in figure 3, the layout of U-shaped high impedance stub of resonator2 increases the effective inductance that enhances the stopband bandwidth of the filter. The optimized structure dimensions are: h_5 = 0.65 mm, l_7 = 4.7 mm, l_8 = 0.45 mm, l_9 = 0.6 mm and w_2 = 0.2 mm. Figures 4 and 5 show the geometry of resonator1 and resonator2 respectively. The open circuited capacitive patches of the resonators are short circuited to the ground at its transmission zero, and so produce a high attenuation level in the stopband. The selected dimensions for patch resonators are: l_{10} = 1.4 mm, l_{11} = 8.2 mm, l_{12} = 1.65 mm, l_{13} = 3.35 mm, h_6 = 1.1 mm, h_7 = 1.6 mm and w_3 = 0.4 mm. The filter analysis is carried out using FR4 material with t = 0.8 mm. The lumped element values of the high impedance stubs as inductance, L and low impedance patches as capacitance, C can be determined by the circuit analysis demonstrated in [8]. Each patch resonator acts as shunt connected series L-C circuits and produces their own transmission zeroes correspond to the resonant frequency f_Z, which can be extracted using (1) as:

$$f_z = \frac{1}{2\pi\sqrt{LC}}. \quad (1)$$

To enhance the stopband bandwidth, resonator3 is introduced with U-shaped high impedance stub and square patches, which serve as suppressing cells at high frequencies.

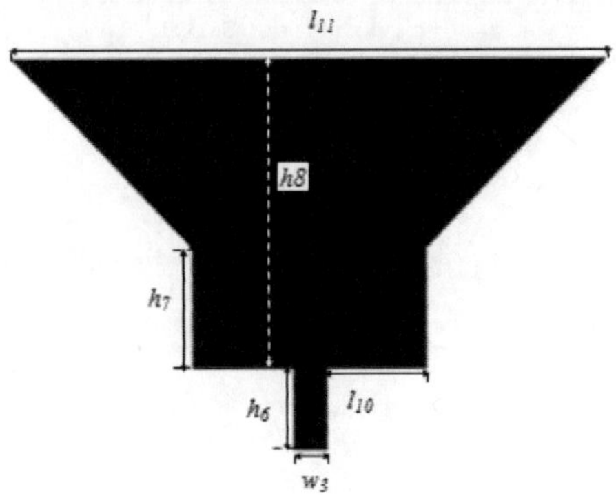

Figure 4. Geometry of modified funnel patch resonator.

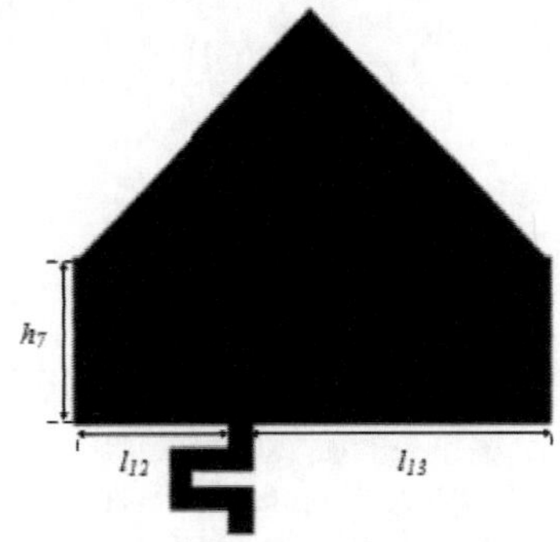

Figure 5. Geometry of modified triangular patch resonator.

3. Simulated and measured results

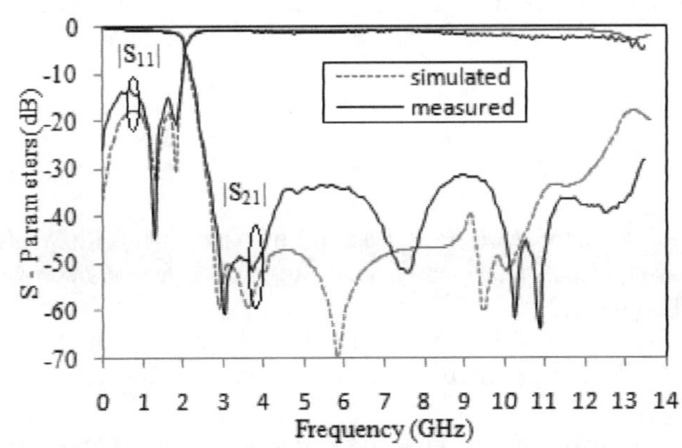

Figure 6. Simulated and measured results of the proposed filter.

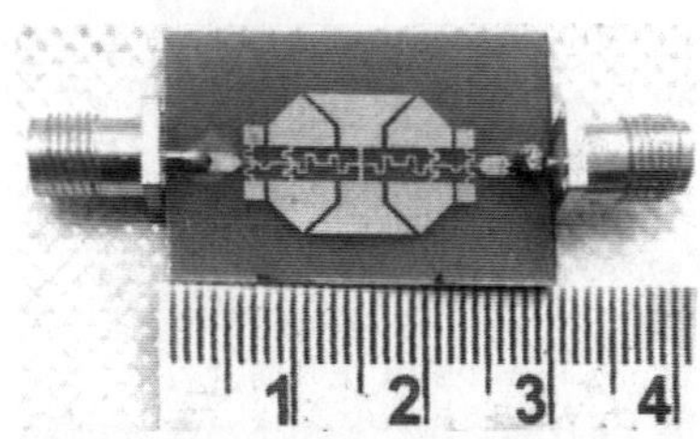

Figure 7. Photograph of prototype filter.

The filter analysis is carried out using FR4 material having permittivity 4.4, thickness 0.8 mm with loss tangent of 0.02. The simulation was accomplished with EM simulation software Zeland IE3D and the measurements are carried out using R&S ZVL 13 Vector Network Analyzer. The simulated results are validated with experimental results as shown in figure 6. The measured 3 dB cut-off frequency of the filter is at 2.04 GHz with a roll off rate of 60 dB/GHz. The filter exhibits high suppression level better than 31 dB with a wide stopband bandwidth from 2.65 GHz to 13.4 GHz. The measured insertion loss of the filter is less than 0.7 dB in the passband at 1.5 GHz. The maximum return loss in the passband of the filter is 13 dB and in the stopband is less than 1.2 dB up to 8.5 GHz and 3 dB in the upper end of the stopband. Excluding the tapered feed line the filter has a compact size of 17.6 mm x 10.6 mm, which corresponds to the normalized circuit size of 0.21824 λ_g x 0.1314 λ_g, where λ_g is the guided wavelength of 80.646 mm at cut-off frequency for 50 Ω impedance. Figure 7 shows the photograph of the fabricated filter. Table 1 presents the performance comparison of the proposed filter with some similar works reported in the literature. As shown in table 1, the proposed filter exhibits compact size and wide stopband with high suppression level among the quoted filters.

Table 1. Comparison between Proposed Filter and Some reported Work.

Ref	fc^a (GHz)	SBW[b] (GHz)	SSL[c] (dB)	NCS[d] (λg^2)	ε_r & t [e] (mm)
[3]	0.5	0.8–4.6	20	0.022	4.4, 1.58
[4]	1.26	1.37–16	20	0.036	3.38, 0.508
[5]	3.12	3.68–19	20	0.056	2.2, 0.787
[6]	5.55	5.58–11.8	15	0.085	4,4, 1.6
[7]	2.5	2.72–10	23	0.052	4.4, 1.6
This work	2.04	2.65–13.4	31	0.029	4.4, 0.8

[a] cut-off frequency
[b] stopband bandwidth
[c] stopband suppression level
[d] normalized circuit size
[e] permittivity and thickness of the substrate material used

4. Conclusion

A compact and wide stopband lowpass filter with high suppression level is designed and demonstrated by loading multiple patch resonators on high impedance corrugated transmission line. The measured results of the proposed filter demonstrate very good electrical characteristics together with compact size, which makes it suitable for practical applications in modern communication systems.

References

[1] Hong J S and Lancaster M J 2001 *Microstrip Filters for RF/ Microwave Applications* (John Wiley, New York)
[2] Ting S W, Tam K W, and Martins R P 2006 Miniaturized Microstrip Lowpass Filter with Wide Stopband Using Double Equilateral U-shaped Defected Ground Structure *IEEE Microwave and Wireless Component Letters* **16** 240–242
[3] Velidi V K and Sanyal S 2011Sharp Roll-off Lowpass Filter with Wide Stopband using Stub-Loaded Coupled-line Hairpin Unit *IEEE Microwave and Wireless Component Letters* **21** 301–303
[4] Karimi G and Lalbaksh A 2013 Design of Sharp Roll-off Lowpass Filter with Ultra Wide Stopband *IEEE Microwave and Wireless Component Letters* **23** 303–305
[5] Hayati M, Sheikhi A, and Lotfi A 2010 Compact Lowpass Filter with Wide Stopband using Modified Semi-Elliptic and Semi-Circular Microstrip Patch Resonator *Electronic Letters* **46** 1507–1509
[6] Raphika P M, Abdulla P, Jasmine P M 2014 Compact Lowpass Filter with a Sharp Roll-off using Patch Resonators *Microwave and Optical Technology Lettersl* **56** 2534–2536
[7] Raphika P M, Abdulla P, Jasmine P M 2015 Planar Lowpass Filter with Sharp Roll-off using Patch Resonators on High Impedance Microstrip Line *IEEE RADIO* 1–2
[8] Ma K and Yeo K S 2003 New Ultra-wide Stopband Low-pass Filter using Transformed Radial Stubs *IEEE Transaction on Microwave Theory and Techniques* **59** 604–611

DC-pass filter design with notch filters superposition for CPW rectenna at low power level

J Rivière, A Douyère, F Alicalapa and J-D Lan Sun Luk

Energy, Electronics and Processes Laboratory (LE2P), University of La Reunion, St Denis, Reunion Island

jerome.riviere@univ-reunion.fr

Abstract. In this paper the challenging coplanar waveguide direct current (DC) pass filter is designed, analysed, fabricated and measured. As the ground plane and the conductive line are etched on the same plane, this technology allows the connection of series and shunt elements to the active devices without via holes through the substrate. Indeed, this study presents the first step in the optimization of a complete rectenna in coplanar waveguide (CPW) technology: key element of a radio frequency (RF) energy harvesting system. The measurement of the proposed filter shows good performance in the rejection of F0=2.45 GHz and F1=4.9 GHz. Additionally, a harmonic balance (HB) simulation of the complete rectenna is performed and shows a maximum RF-to-DC conversion efficiency of 37% with the studied DC-pass filter for an input power of 10 μW at 2.45 GHz.

1. Introduction

Energy harvesting has been the center of attention recently with the development of Internet of Things [1]. Energy can be harvested from different sources, such as the sun, the wind, vibration and electromagnetic waves. Moreover, with the increase of wireless data transfer in our environment, electromagnetic energy can be considered as a source of energy. This energy could be collected and used to power small devices, such as sensors [2]. This is achieved, using rectenna (Figure 1.a). One of the challenges relies on the repeatability of their conception. Specifically the repeatability of vias needed to connect surface mounted component (SMC) to the ground. Rectenna realizations should be robust to improve repeatability. This is important, because energy harvesting often requires complex rectenna or more than one rectenna due to the low level power of available electromagnetic energy. Hence, antenna array or rectenna array are used to provide higher voltage to power the storage energy unit or the device itself. The first solution is convenient when dealing with simple structure and mono frequency antenna. But, as the complexity of antenna and number of frequency increase, one has to deal with phases shifting and complex matching impedance circuitry. The second solution, rectenna array, avoids those problems by combining the DC voltage of several rectenna. However, this method requires many vias in microstrip technology that can lead to performance shifting in rectenna [3].

This work is dedicated to the second possibility. To limit risks of frequency shifting or performances degradation, we avoid the use of vias. To achieve this, we use coplanar waveguide (CPW) (Figure 1.b) instead of microstrip. As the ground plane and the conductive line are etched on the same plane, drills in the substrate are no longer required. The topic of this paper focuses on the design and conception of an easy-to-build, compact and uniplanar DC-pass filter for CPW rectenna.

First, two discrete-element notch filters are simulated and their equivalent circuits in CPW technology are conceived and tuned with Momentum in Agilent ADS software. Then, a DC-pass filter is realized by superposition of the two notch filters.

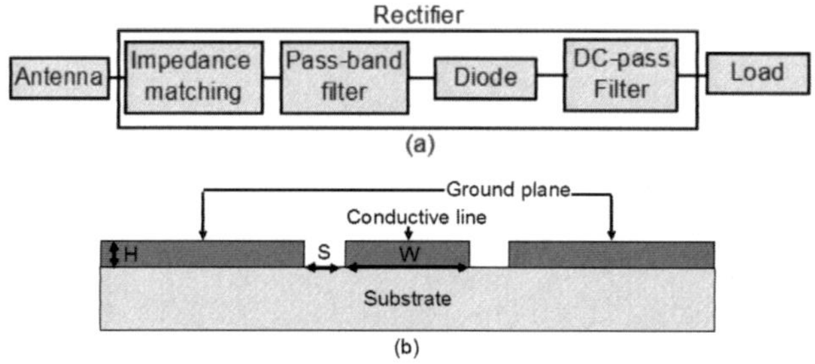

Figure 1. Synoptic of a rectenna(a) and CPW structure (b)

2. DC-pass filter design

To improve rectifying efficiency, one must ensure proper DC voltage to the load. This can be achieved by using a parallel capacitor as DC-pass filter after the rectifying process of the diode. In the literature, the equivalent model of a parallel capacitor in CPW can be modeled by an open-end coplanar shunt stub in the ground plane [4]. But, this solution cannot be used in our rectenna conception because it implies the use of an air bridge to ensure equality of the potential ground. Indeed when a discontinuity is etched in the CPW circuitry, the propagation of the parasitic coupled slot line mode should be considered.

The investigation of other filters leads to the following structure: instead of the well-known parallel capacitor, a series inductance or a resonant structure can act as a DC-pass filter.

Subsequently to this work, our final objective is to realize via-less rectenna working in the ISM band, and precisely at F0=2.45 GHz to use in a direct RF source feeding scheme. In rectenna circuits, rectifying process of the diode generate higher order harmonics (F0,F1=4.9 GHz, etc) in addition to the DC. To achieve higher conversion efficiency, those harmonics must be cancelled out from the DC output. Working at low input power level, we considered the effects of higher order harmonics superior to F1 negligible. Hence, the conception of only two notch filters is performed at F0 and F1, which in this case when combined together act as a DC-pass filter.

2.1. 2.45-GHz and 4.9-GHz notch filter

The first proposed filter is realized with a series short stub in the conductive line on AD1000 substrate, with $\varepsilon r=10.35$, substrate height h=0.762 mm and dielectric loss $\delta=0{,}002$. The stub length value, Ls0=12.48 mm, is obtained through tuning using Momentum simulator. Concerning the etching process steps, all the CPW notch filter slots have been done with the same width for simplicity (S=0.7874 mm). Hence, the conductive line width W=3.7 mm to obtain a characteristic impedance line of 50 Ω. This structure does not introduce any gap in the conductive line, so the DC part of the signal can reach the load. The simulated planar filter S-parameters and the discrete-element equivalent circuit are plotted in Figure 2.

Using the same principle, a second resonant structure is designed in order to reject the first higher order harmonic at F1=4.9 GHz corresponding to Ls1 (Figure 2.b). This structure is represented in CPW by a series stub in the conductive line with length Ls1=5.98 mm.

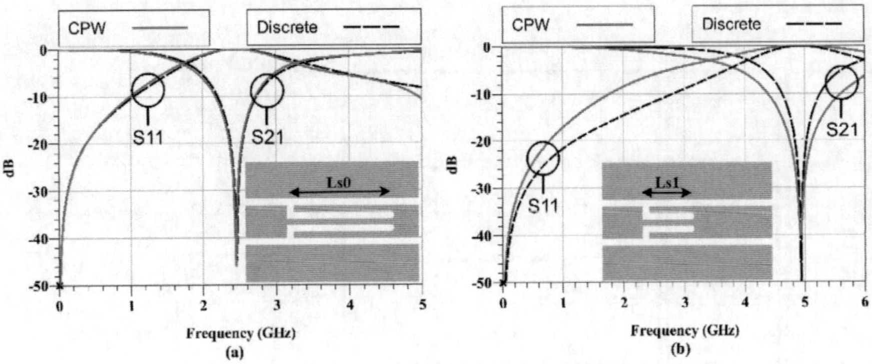

Figure 2. S-parameters of the discrete-element and CPW simulated 2.45-GHz notch filter (a) and 4.9-GHz notch filter (b) with their respective layout representation.

2.2. *The complete DC-pass filter*

Both of the two notch filters are etched on the conductive line of the CPW structure. In order to create the DC-pass filter, the two frequencies must be filtered. One could combine them in series to achieve the global DC-pass filter. But, this type of combination increases the size of the global DC-pass filter. Another solution when realizing CPW stub in series, is to etch the stub in the ground plane, which shows similarity with defected ground structure (DGS). But, by etching the stub in the ground plane, the bandwidth is narrowed for the chosen frequency, leading to better quality factor. In our study, we know that the first higher order is predominant over the second one in the output signal. Consequently, the 2.45-GHz notch filter was chosen to be milled on the ground plane with a new slightly modified value, Ls0'=13.5 mm. Using the superposition principle, both filters are etched in parallel on the same board (Figure 3).

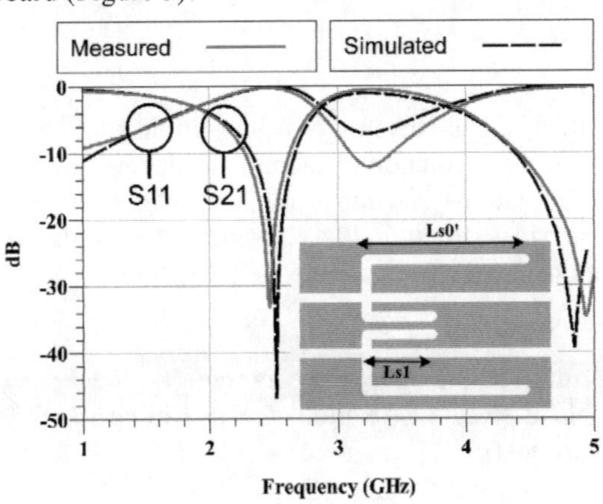

Figure 3. S-parameters of the measured and simulated proposed DC-pass filter with a layout representation

3. Measurement and application

The proposed DC-pass filter has been fabricated and measured (Figure 3). The measured and simulated results show good agreements. But a frequency shift is observable in Figure 3. We believe this shift is explained by the milling process and the precision in substrate dielectric value $\Delta\varepsilon r=\pm 0.35$.

Subsequently to this work, a first rectenna (Figure 4) using the proposed DC-pass filter and a discrete-element matching circuit, has been simulated with ADS. The S1P (1-port) block represents the S parameters of a CPW dipole antenna and the S3P (3-ports) block allows the separation of the injected power and the reflected power. The matching circuit is composed of an inductance L4=9.9 nH and a capacitor C5=5.2 pF. Results shows a rectifying efficiency up to 37% at 2.45 GHz for Pin=-20 dBm (Figure 5).

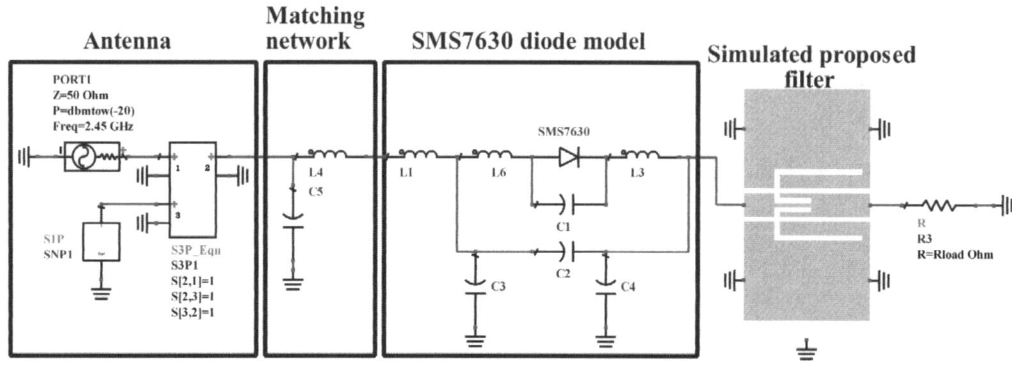

Figure 4. Schematic view of the simulated rectenna at -20 dBm, 2.45 GHz and a load R=5400 Ω

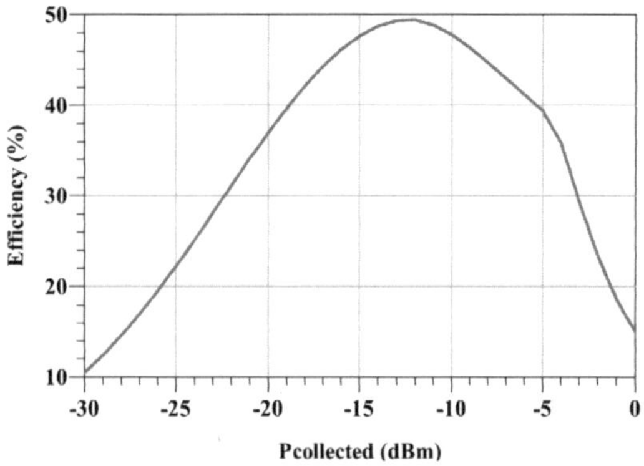

Figure 5. RF-to-DC conversion efficiency of the simulated rectenna with the proposed DC-pass filter.

4. Conclusion

In this paper, using the superposition of two notch filters, a DC-pass filter with multiple stop bands was realized and characterized on CPW. We showed that the prediction of the notch filters is possible with a simple resonant LC model. The Momentum simulations and measurements of the DC-pass filter showed good agreements. Thanks to the CPW physical structure, this DC-pass filter is suitable for a via-less rectenna design and its principle was demonstrated within a rectenna simulation.

Acknowledgments

We acknowledge the financial support of structural funds of the European Community, the French Government and the Regional Council of Reunion Island (Region Reunion) for providing research grants (POFEDER PRESAGE 2013-32 933 CARERC Project).

References

[1] Perera C, Zaslavsky A, Christen P and Georgakopoulos *D 2014 Context aware computing for the internet of things: A survey* IEEE communications survey & tutorials vol 16

[2] Mitcheson P D, Yeatman E M, Kondala Rao G, Holmes A S and Green T C 2008 *Energy harvesting from Human and machine motion for wireless electronic devices* Proc. of the IEEE vol 96

[3] Rivière S, Douyère A, Alicalapa F et Lan Sun Luk J D 2011 *Synthèse d'un module énergétique en vue d'alimenter à distance des dispositifs électroniques* Conf. Journées Nationales Microondes, (BREST)

[4] Simons R N 2001 *Coplanar waveguide circuits, components, and systems* chapter 9 ed Wiley IEEE press

Simulations and Tests of Prototype Antenna System for Low Frequency Radio Experiment (LORE) Space Payload for Space Weather Observations

Kaiwalya Pethe[1], Shridhar Galande[1], Sachin Jamadar[1], S. P. Mahajan[1], R. A. Patil[1], B. C. Joshi[2], P. K. Manoharan[2,3], Jayashree Roy[2], G. Kate[2]

[1] Department Of Electronics and Tele-Communication, College of Engineering, Pune, India
[2] National Centre for Radio Astrophysics (TIFR), Pune, India
[3] Radio Astronomy Centre, National Centre for Radio Astrophysics (TIFR), Ooty, India

E-mail: pethekaiwalya@gmail.com

Abstract. Low frequency Radio Experiment (LORE) is a proposed space payload for space weather observations from space, operating between few kHz to 30 MHz. This paper presents preliminary design and practical implementation of LORE antenna systems, which consist of three mutually orthogonal mono-poles. Detailed computational electromagnetic simulations, carried out to study the performance of the antenna systems, are presented followed up by laboratory tests of the antennas as well as radiation tests with a long range test range, designed for this purpose. These tests form the first phase of the design and implementation of the full LORE prototype later in the year.

1. Introduction

Low frequency Radio Experiment (LORE) is a proposed space payload for high-time resolution and high-frequency resolution space weather observations from space, operating between few kHz to 30 MHz. Its main scientific objective is to study the propagation of coronal mass ejections (CMEs) from the Sun using Type II and Type III bursts as probes in conjunction with ground based facilities, such as Ooty Radio Telescope (ORT), which are already part of the space weather monitoring program. The proposed design consists of three mutually orthogonal mono-pole antennas of 5-m length mounted on one side of spacecraft. Digitized signals from these antennas, sampled at 1 ms time resolution and 30 kHz or more frequency resolution, are proposed to be correlated on-board to perform gonio-polarimetry, which will allow simultaneous direction finding and polarization measurements for Type II and Type III bursts during solar energetic events.

2. Antenna System Design and Simulations

In this project, simulations and tests were carried out for a scaled design at 70.9 MHz to validate the simulation results, primarily because we have license for radiation tests at this frequency at Ooty. The design is later proposed to be scaled to 0–30 MHz for the actual payload. In the 70-MHz design, computational electromagnetic simulations, using Computer Aided Design

Figure 1: The 70-MHz monopole antenna configuration with a ground-plane, ready for experiments at the test range at the Radio Astronomy Centre, Udhagamandalam.

package (CADFEKO)[1] were carried out to get as uniform a sensitivity as possible from the antenna system, a reasonable directional beam and as high gain as possible. The proposed antenna configuration for LORE to carry out direction finding and polarization measurements simultaneously is pictorially shown in Figure 1 and it is similar to that used in a previous solar mission [1].

Figure 2 shows the 3-dimensional simulated beam pattern for the system with excitation to one mono-pole and the ground plane. The simulations were verified against test results (refer to Section 3) conducted in a test range designed by us. Due to large size on the antenna configuration, measurement of the beam pattern was possible in the azimuthal coordinate only. A comparison between the simulated beam pattern and test results in the azimuthal coordinate for the system with excitation to one monopole and the ground plane is shown in Figure 3. The test result is consistent with the simulated beam pattern (with a reduced chi-square value of $\sim$3). However, the simulated and practical radiation patterns are slightly different, which is probably due to the finite ground plane, lossy antenna electrical connections, pickups from human body, small antenna height (1.2 m) from ground and so on.

Figure 4 shows the simulated beam for all the three monopoles excited as a phased array. As an alternative to ground plane, simulation for one excited mono-pole with three parasitic *star-shaped* mono-poles, was also carried out and tested in practice. We have simulated the model for excitations with varying phase difference demonstrating that gonio-polarimetry can be performed by combining different directional information from three monopole antennas with proper phase shift.

The antennas, optimized after simulations, were then fabricated at NCRA-TIFR and these

[1] *https://www.feko.info*

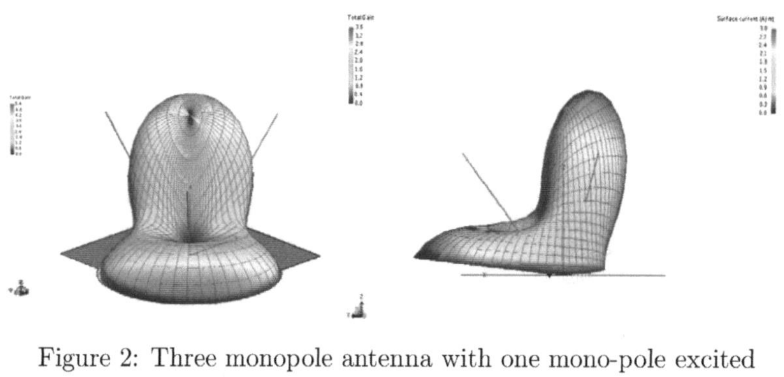

Figure 2: Three monopole antenna with one mono-pole excited

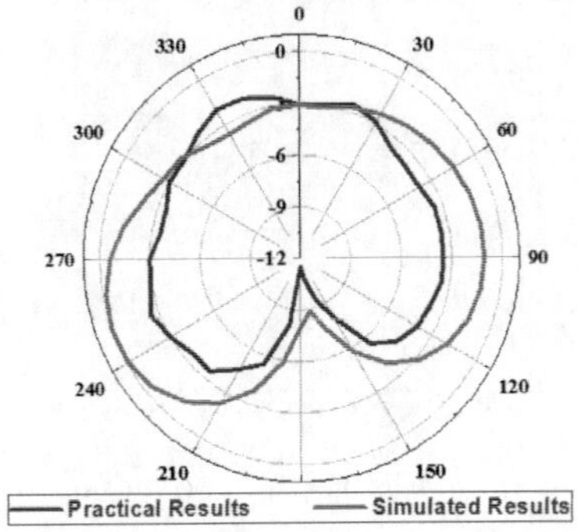

Figure 3: Radiation pattern in the azimuthal plane when only one monopole antenna is excited

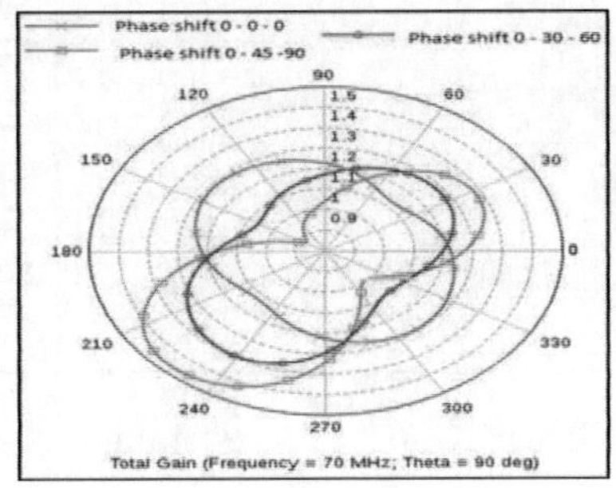

Figure 4: Beam pattern simulation with various phase difference values between antennas.

practical implementations were tested with a vector network analyser (VNA). The results obtained in VNA tests are also plotted in Figure 5 along-with the simulated results and they show a close match validating the simulations.

3. Field Tests

Field trials with a temporary test range, designed by us, were carried out at the Radio Astronomy Centre, Udhagamandalam, to validate the other results from the simulations. A test range consists of instruments like antennas, transmitting and receiving systems, positioning system and recording system. There are different types of ranges such as the elevated range, the slant range, the compact range, and test ranges in an anechoic chambers. We used a modified version of slant and elevated range for our field tests [2].

The location chosen was Udhagamandalam (Ooty) as the hills in Ooty naturally provide elevation to avoid ground reflections and field tests can be carried out over several kilometres due to large distance between the hills. The radiating antenna in the test range was put on top of a hill at Kollaribeta, whereas the receiving antenna was located near the Radio Astronomy Centre of NCRA-TIFR at Ooty. The distance between these two stations is approximately 10 km. The schematic of the test range is shown in Figure 6

A signal generator with 20 dBm power output was used for excitation, whereas a hand-held spectrum analyzer with a noise floor of -80 dBm was used for measurement of received signal. We simulated, fabricated and tested several types of antennas for our test range, namely a simple dipole, dipole with a reflector and a Yagi-Uda antenna [3]. The comparison of antenna parameters from the simulation of these different types of antennas using CAD-FEKO is shown

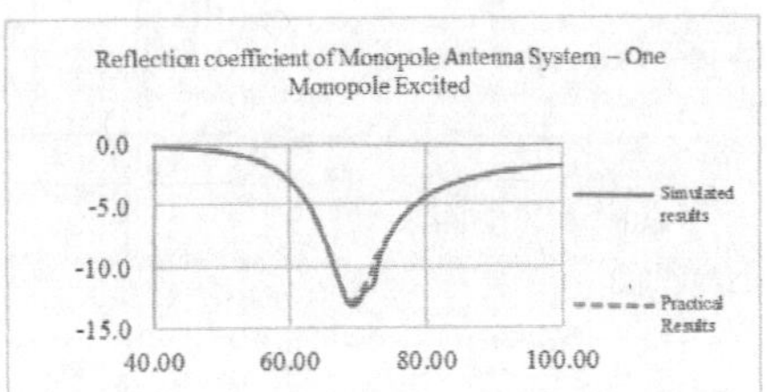

Figure 5: Reflection coefficient of monopole antenna system.

Antenna	Antenna Impedance(Ohm)	Reflection Coefficient (dB)	Gain	3-dB Beam width (Azimuth Plane)	-10 dB Bandwidth (MHz)
Half-wave Dipole	69.64	-15.51	1.63	78.75	5.58
Reflective Dipole	42.90	-20.75	4.41	69.22	4.171
4 Element Yagi-Uda Antenna	51.05	-39.03	7.39	60.67	7.55

Table 1: Parametric Chart of the tested antennas

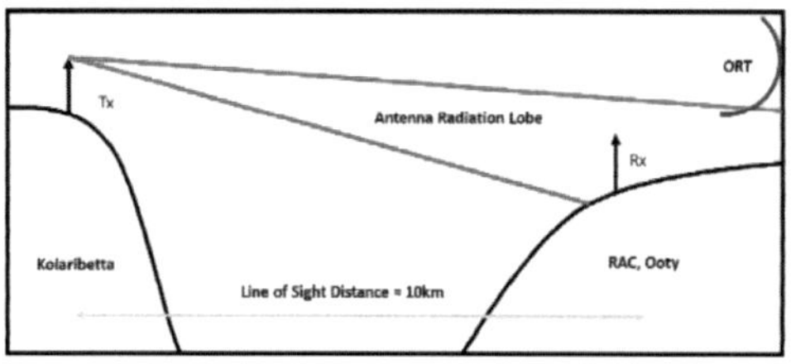

Figure 6: Antenna test range schematic.

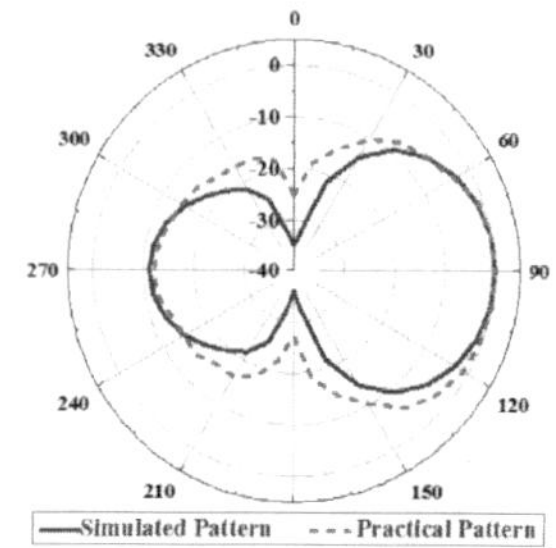

Figure 7: Radiation pattern of the Yagi-Uda antenna.

in Table 1. Yagi-Uda antenna was eventually chosen for field test due to its high directivity and bandwidth. All antennas were adjusted for an input impedance of 50 ohms. First Yagi-Uda antennas were placed on both transmitting and receiving end and their beam pattern were obtained by rotating the receiving element with 10 degree steps. These are shown in Figure 7, along with the results from CAD-FEKO simulations. The obtained pattern was used in the deconvolution to obtain the pattern for the payload antenna systems. Then, the three-monopole antenna system was tested by keeping the source antenna was as Yagi-Uda antenna and the monopole antenna system as receiver.

In order to calculate the power received or the gain of the antennas, we use Friis Transmission Equation, which is,

$$P_r = P_t + G_t + G_r + 20\ \log_{10}(\frac{\lambda}{4\pi \mathrm{R}}) - \mathrm{loss} \qquad (1)$$

where G_t and G_r are the antenna gains (with respect to an isotropic radiator) of the transmitting and receiving antennas respectively, λ is the wavelength, and R is the distance between the antennas. From the experimentally measured values (P_r = -56.5 dBm, P_t = 20 dBm, G_t = 7.39 dB, $R \approx 10$ km, and $\lambda = 4.28$ m), we obtain $G_r = 5.5$ dB, for the monopole configuration, proposed to be used for LORE.

4. Conclusions and Future Work

Satisfactory match between the simulations and field tests were obtained for the proposed three monopole system for LORE, with antenna scaled to 70.9 MHz for ease of testing. This provides confidence in design for 0–30 MHz antenna systems with the computational electromagnetic platform. This is important as good field tests at this frequency range will be difficult and challenging and reliance on computational electromagnetic simulation is critical. The design for 0–30 MHz antenna configuration is currently in progress. The monopole design is well suited for space applications as monopole with high stiffness can be made from rolled shape memory alloy providing a very low stow volume and low weight. Nevertheless, alternative 0–30 MHz antenna designs are also being explored.

Acknowledgments

This work utilized support from XII plan grant TIFR-12P0716.

References

[1] Bale S D, Ullrich R, Goetz K, et al.2008 *Space Science Reviews*, **136** 529
[2] *IEEE Standard Test Procedures for Antennas*, New York: IEEE, Std. 149-1979
[3] Balanis C A, *Antenna Theory Analysis and Design*, 2005, 3rd ed., John Wiley

Hyperband Bi-Conical Antenna Design Using 3D Printing Technique

J.A. Andriambeloson, P.G. Wiid

Electrical & Electronic Engineering Department,University of Stellenbosch, Western Cape, South Africa

E-mail: andyjoely@gmail.com, wiidg@sun.ac.za

Abstract. We combined a 3D printing technique with conductive paint for an antenna manufacturing methodology. The performance of the approach is evaluated through a 3D-printed and coated bi-cone antenna. The antenna far-field pattern and efficiency are measured using near-field spherical scan and reverberation chamber techniques. Good agreement is seen between measurements and simulations and an impedance bandwidth of at least 34:1 is achieved. An extruded bi-conical antenna geometry is also studied for bandwidth extension to lower frequency and an impedance bandwidth of 58:1 is realised.

1. Introduction

Over the past few years, the process of making objects with complex shapes has become easier due to the development and accessibility of open-source 3D-printing machines to the general public. The new tool not only allows an accurate replication of numerical models that are drawn on a computer, but it also permits a light-weight and a low-cost design using plastic materials such as a polylactic-acid (PLA). Here, it is combined with the HSF55 carbon-based conductive paint for low-cost antenna design methodology evaluation. A bi-conical 3D-printed omni-directional antenna is built and is examined to understand the performance of the approach. The study focuses on the antenna under-test (AUT) matching, far-field and efficiency responses. Anechoic and reverberation chambers, as well as computational models are used for validation of the measured antenna parameters.

2. Antenna Under-Test Description

Figure 1.a shows a photograph of our open-source 3D printer printing the truncated cone that was utilised to manufacture the bi-cone antenna in Figure 1.b. The plastic cone is printed with an infill density of 15% using a PLA plastic. A conical machined-piece of aluminium is inserted on top of each truncated-cone (Figure 1.c) and the assembled product is coated with a carbon-based conductive paint: HSF55. The bi-cone is characterised by an angle of 41 degrees for optimal impedance matching over a wide frequency band [1]. It is fed from the bottom of the lower cone through a coaxial cable of 20 cm long as described in Figure 1.c.

3. Antenna Parameters Measurement and Validation

Three antenna parameters are investigated: the bi-cone reflection coefficient (S_{11}), the electric far-field directivity (EFD) pattern and the antenna efficiency. A spherical near-field scan is used

IOP Conf. Series: Materials Science and Engineering **120** (2016) 012010 doi:10.1088/1757-899X/120/1/012010

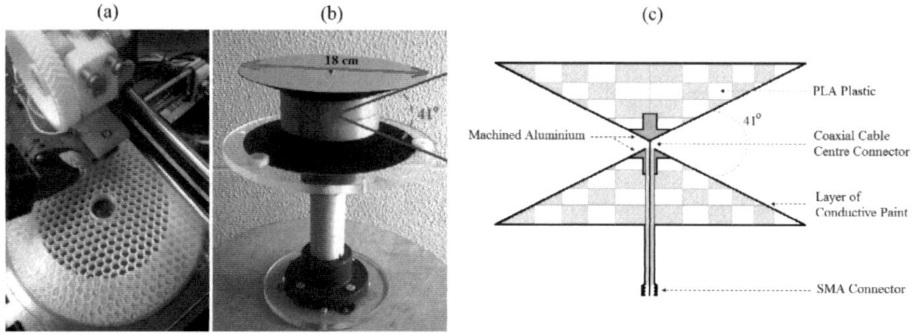

Figure 1. Overview of the 3D-printed bi-cone antenna manufacturing: (a) shows the 3D-printer, (b) displays the finished bi-cone antenna and in (c) is described the antenna feed.

for the radiation pattern investigation while a reverberation chamber technique is employed for the efficiency characterisation. The bi-cone is placed inside our anechoic chamber within the AUT radiative near-field region for the EFD measurement. The AUT normalised EFD pattern is extracted from the spherical near-field data using our anechoic chamber software: NSI2000. As far as the efficiency measurement is concerned, the bi-cone radiation efficiency is derived from the efficiency of a known dipole of 10 cm long that we characterised with FEKO in previous work [2]. Two setup scenarios (S1 and S2) are performed for the investigation and for both, the same radiating antenna is successively placed at the eight corners of the chamber working volume. The power radiated within the chamber is recorded with a vector network analyser (VNA) through the reference-dipole antenna (located at a fixed place) for S1 and through the bi-cone for S2. 72 stirrer steps per revolution are used and the bi-cone efficiency is calculated using the average of the 72×8 S_{21} matrix collection as expressed in (1). It is also important to notice that here each S_{21} is corrected from port mismatches.

$$\eta_{Cone} = \eta_{Dipole} \frac{< |S_{21\,(S1)}|^2 >}{< |S_{21\,(S2)}|^2 >} \tag{1}$$

For the bi-cone simulation, the paint coating was modelled using a uniform layer of perfect electric conductor (PEC), aluminium and the CST predefined graphite material. The PEC provides the idealised antenna response while the aluminium and the graphite allows us to inspect the variation of the antenna radiation when the coating layer conductivity changes.

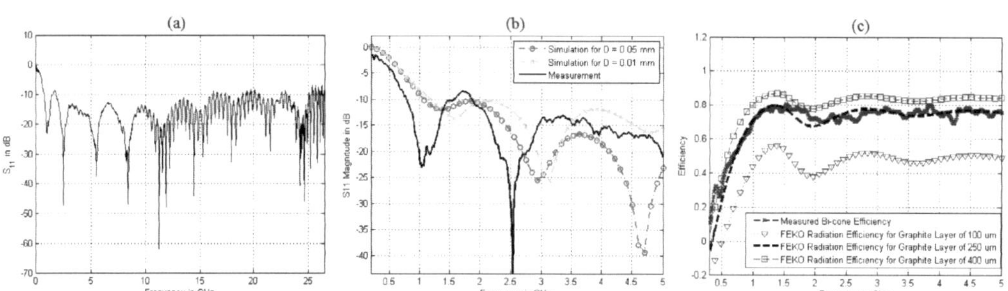

Figure 2. Reflection coefficient of the bi-cone antenna: (a) shows the measured S_{11} using a 26.5 GHz Agilent PNA-X VNA and in (b) it is compared with simulated curves up to 5 GHz. (c) displays the measured and the simulated antenna radiation efficiencies.

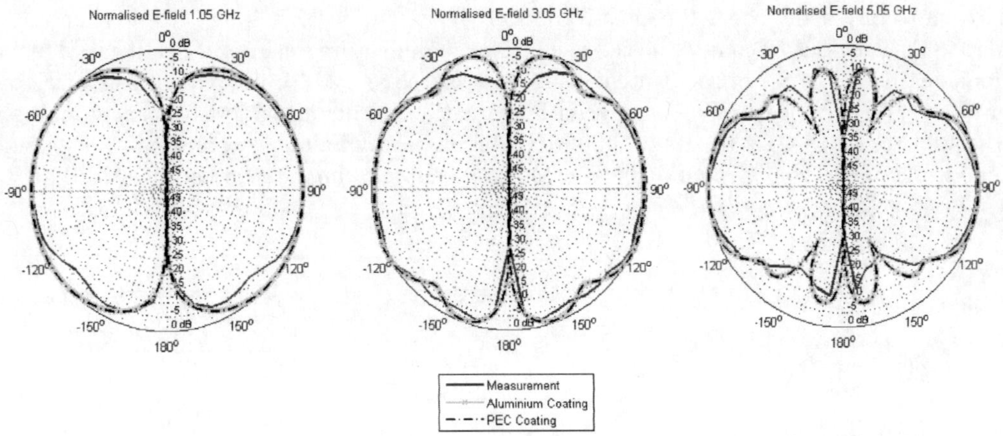

Figure 3. Comparison between measured and simulated bi-cone normalised EFD for 1.05 GHz, 3.05 GHz and 5.05 GHz.

4. Results and Discussion

Figure 2.a shows the antenna S_{11} as measured with a 26.5 GHz Agilent PNA-X VNA. Using a matching threshold of -10 dB, the bi-cone achieved an impedance bandwidth of at least 34:1. The measured S_{11} is compared with simulation results up to 5 GHz on Figure 2.b. Here, a good agreement is seen between the curves in terms of shape and level, but different resonances are observed. This behaviour is expected since the antenna S_{11} is influenced by the gap spacing (D) between the upper and the lower cones as found in [1]. The model coating thickness has been adjusted to fit the measured efficiency result and a graphite layer of 250 m thick was found to be the best match for the physical antenna coating (see Figure 2.c).

Figure 3 compares the measured and the simulated normalised EFD of the bi-cone antenna for 1.05 GHz, 3.05 GHz and 5.05 GHz. The bi-cone main-lobes agree with computations between -60 and 60 degrees, but clear deviations are observed within the side-lobe regions located around 0 and 180 degree. The overall simulated far-fields are symmetric relative to the horizontal line of the polar axis, but no symmetry is observed for the measured curves coloured in blue. This observation is clearly shown by the curves for 1.05 GHz and this is due to an unbalanced feed problem where common-mode currents flow over the feed-cable external conductor and radiate unwanted energy in the antenna surrounding. The unbalanced-feed issue can be solved using a balun or a differential amplifier, but these solutions are not appropriate for the bi-cone due to its hyperband response. A different approach is necessary to reduce the feed-cable effect and this will be investigated in further work.

5. Antenna Response Improvement for Low-Frequency Signals

The low-frequency limit of the bi-cone antenna can be extended using a larger cone-radius. However, our 3D-printer cannot print a conical structure having a radius more than 9 cm. Another technique had to be investigated and here, we used an approach which is similar to the one presented in [3]. We extruded the base of each cone by a height H to allow the bi-cone to act as a dipole antenna at low-frequency (see Figure 4.a). Four FEKO models were used for the study where the height of the cylindrical expansion (H) was varied from 4 cm, twice the cone height L, 8 cm and 10 cm. The simulated S_{11} are shown in Figure 4.b. We can see here that the extrusion of 8 cm provides the lowest cut-off frequency and where the antenna S_{11} remains less than -10 dB from around 450 MHz. Two new bi-cones have been realised using a cone-radius of 8.57 cm to physically check the effectiveness of the extrusion technique. The first antenna is similar to the previous bi-cone while the second was manufactured using the extrusion value of 8 cm found earlier. The S_{11} of both antennas were measured and are compared with simulations in Figure 4.c. The two measured S_{11} agree with simulation results and a low-frequency limit shift is observed from 1.18 GHz to around 450 MHz due to the cylindrical expansion. The extruded bi-cone works up to 26.5 GHz and this increases the antenna impedance bandwidth to at least 58:1.

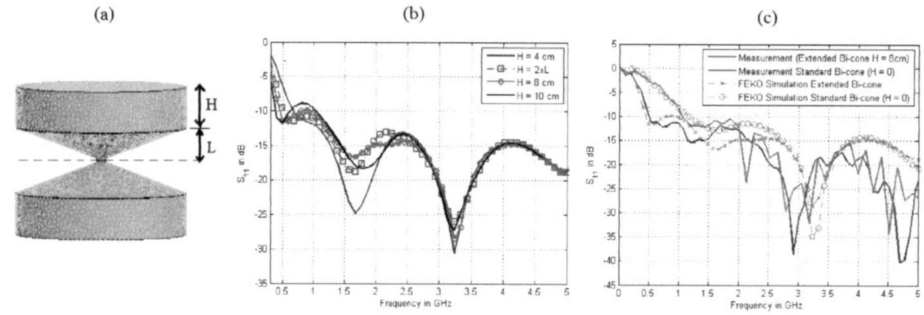

Figure 4. Description of the bi-cone extrusion technique for low-frequency limit improvement. (a) and (b) show the computational model and the simulated S_{11} for four extrusion values. (c) compares our measurement results with simulation using a cylindrical expansion of 8 cm.

6. Conclusion

With low-cost in mind, a 3D-printed bi-cone antenna was manufactured to investigate the performance of a 3D-printing technique and conductive paint for an antenna manufacturing methodology. The antenna S_{11}, electric far-field pattern and efficiencies were measured, using anechoic and reverberation chambers, and obtained results were validated using simulations. A uniform graphite layer was used to model the antenna coating layer and this provided a good agreement with measurement results. The current antenna design achieved a -10 dB impedance bandwidth ratio of 34:1 (770 MHz to 26.5 GHz). An efficiency of more than 70% was measured with the reverberation chamber technique up to 5 GHz. The low frequency cut-off of the bi-cone was also improved using a cylindrical extrusion technique and our current prototype achieved an impedance bandwidth of 58:1. The study shows the effectiveness of the 3D printing technique and the HSF55 conductive paint for antenna design, but further work is necessary to solve the antenna unbalanced feed issue. In this regard, a combination of L-plates and absorbers will be investigated to attenuate the flow of common-current across the feed-cable external conductor.

Acknowledgement

We acknowledge the following people for their help in the work: Dr H. Pienaar, Mr W. Croukamp, Ms A. Bester, MeerKAT team, Altair-FEKO and CST, SKA-South Africa and SARCHI for funding.

References

[1] Wiid P. G. *The Answer is in Fact 41, or How to Get 35:1 Bandwidth from a Cone Antenna*, *Int. Conf. on Antenna and Propagation in Wireless Communications (APWC)*, pp 1060-1063,Turin, Italy, 7-11 Sept 2015.
[2] Andriambeloson J. A. *Reverberation Chamber Time and Frequency Metrology for MeerKAT Systems Shielding Evaluation*, Chap 3, pp.26-42, Dec 2014, http://hdl.handle.net/10019.1/96037.
[3] Keller S. D. *Wire-Frame Monocone Antenna for Direction-Finding Applications on Unmanned Aerial Vehicle Platform*, *Antennas and Propagation Magazine*, IEEE, vol.53, no.1, pp.56,65, Feb. 2011.

Investigation of spherical and cylindrical Luneburg lens antennas by the Green's function method

S. Knyazev, A. Korotkov, B. Panchenko and S. Shabunin

Ural Federal University, Ekaterinburg, Russia

E-mail: shab@ieee.org

Abstract. Luneburg lens antenna radiation fields are calculated with Green's functions of spherical and cylindrical layered structures. Electric field components of spherical and cylindrical Luneburg lenses excited by linear and circular polarized incident field are analysed. Dipole, slot and aperture antennas are described by electric and magnetic extraneous currents. Radiation patterns of cylindrical and spherical Luneburg lens are analysed. Co-polarized and cross-polarized field radiation patterns are shown. The proposed method significantly reduces the computing time for multi-layered lenses in comparison with the most commonly used in antenna design. The first step antenna structure optimization may be performed for a shorter time. The results may be used as the first approximation for Ansys HFSS and other software.

1. Introduction

Spherical and cylindrical Luneburg lenses are suitable for multiple-beam and scanning high-gain antenna systems design. The refraction index of Luneburg lens varies as follows [1]

$$n(r)=\sqrt{\varepsilon'(r)}=\sqrt{2-(r/a)^2}, \qquad (1)$$

where ε' is the relative dielectric constant of lens material at the point r, r is the radial coordinate in the spherical or cylindrical coordinate system, a is the outer lens radius in the corresponding coordinate system. Usually Luneburg lens is fabricated as a multilayer structure. The refractive index of the each layer is close enough to the law of Luneburg as in equation (1).

There are a lot of papers devoted to Luneburg lens antennas analysis with modern electromagnetic software as Ansys HFSS [2], CST Microwave Studio [3] and others. The authors have made a note of more processing time for simulation. The optimization of the complicated antenna system as the Luneburg lens with a feeder is a multivariate optimization problem. The initial solution is close to the optimum significantly accelerating calculations. Our goal is to offer a new method for the first step antenna structure approximation.

We suggest the well-known Green's function method for the first step Luneburg lens antenna analysis and optimization. This method gives a chance for physical interpretation of obtained results but it is rather limited in the shape and structure of qualified objects. Fortunately, Luneburg lens may be well fitted by the spherical or cylindrical coordinate system. The antenna radiation field is defined by integrating of electric and/or magnetic extraneous currents with the Green's functions suitable for the spherical or cylindrical coordinate system. As extraneous currents electric or/and magnetic dipoles are used. For example crossed electric and magnetic dipoles form the Huygens source. This kind of the source is used for aperture antennas design as a horn antenna. A linear waveguide with slots for cylindrical Luneburg lens excitation is described by magnetic dipoles.

In this paper a universal algorithm with the Green's functions of layered spherical and cylindrical structures for Luneburg lens antenna radiation problems solving is described. The offered method demands at least one thousand less time for calculations in comparison with Ansys HFSS. Radiation patterns, antenna gain, directivity, radiation efficiency and polarization properties may be rather quickly analyzed.

2. Spectral-domain full-wave approach for multi-layered spherical and cylindrical structures

The spectral-domain full-wave approach suggested in [4],[5] uses the rigorous solution of the Maxwell equations for radially non-uniform structures with arbitrary electrodynamic properties and any distribution of impressed electrical and magnetic currents. This method with matrix approach is suitable for MATLAB. It is easy to simulate layered structures, impedance surfaces, and coating layers including metamaterials.

According to the suggested method the main electric and magnetic field component is chosen. Other components are defined with the main components. In the spherical coordinate system as the main radial electric E_r and magnetic H_r components are used. In cylindrical coordinates E_z and H_z are the main and the first one may be represented as

$$E_z(r,\varphi,z)=-\frac{1}{4\pi^2}\sum_{m=-\infty}^{\infty}\int_{V'}\int_{-\infty}^{\infty}\left[\Phi_1^E(m,h;r,r')+\Phi_2^E(m,h;r,r')\right]\exp\left[-jm(\varphi-\varphi')\right]\exp\left[-jh(z-z')\right]dh\,dv', \quad (2)$$

where functions Φ_1^E, Φ_2^E contain information about sources, cylindrical structure dimensions and electrodynamic material properties. These functions may be considered as the Green's functions. In equation (2) all primed symbols are source points. The integration is carried out in relation to the source region V'.

We can calculate electromagnetic field from any point of view. To obtain antenna radiation patterns the saddle-point method is used. For any point of the space the electric field components in the spherical coordinate system are defined as follows

$$E_\theta(\theta,\varphi,r)=\frac{1}{2\pi^2}\sum_{m=0}^{\infty}\varepsilon_m\cos(m\varphi)\,j^m\frac{\exp(-ik_0r)}{k_0r}\left(\Phi_1^E+\Phi_2^E\right)\left(\frac{\cos^2\theta}{k_0\sin^2\theta}+\sin\theta\right), \quad (3)$$

$$E_\varphi(\theta,\varphi,r)=-\frac{1}{2\pi^2}\sum_{m=0}^{\infty}\varepsilon_m\cos(m\varphi)\,j^m\frac{\exp(-ik_0r)}{k_0r}\left(\Phi_1^M+\Phi_2^M\right)\frac{Z_0}{k_0\sin^2\theta}, \quad (4)$$

where $\varepsilon_0=1,\ \ \varepsilon_m=2,\ m\neq 1$.

Electric field components of a spherical Luneburg lens antenna in the far zone are simplified by saddle-point method in the spherical coordinate system. Electric field is defined as follows

$$\mathbf{E}(\theta,\varphi,r)=E_0\frac{\exp(-ik_0r)}{k_0r}\sum_{n=1}^{\infty}\frac{2n+1}{n(n+1)}\exp\left[i\left(\frac{\pi}{2}n+\varphi\right)\right]\left\langle\begin{matrix}[\mathbf{a}_\theta\tau_n(\theta)+i\mathbf{a}_\varphi\pi_n(\theta)](M'_n+jM''_n)-\\-[\mathbf{a}_\theta\pi_n(\theta)+i\mathbf{a}_\varphi\tau_n(\theta)](N'_n+jN''_n)\end{matrix}\right\rangle, \quad (5)$$

where $k_0=2\pi/\lambda$ is the wave number, $\mathbf{a}_\theta,\mathbf{a}_\varphi$ are unit vectors in the spherical coordinate system, $\tau_n(\theta)$ and $\pi_n(\theta)$ contain Legendre functions and derivatives of these function. Coefficients M_n, N_n in (5) depend on an excitation current and layered structure properties.

Luneburg lens with different kind of feed may be simulated at the corresponding choice of amplitude and phase of extraneous currents.

For radially layered structure, a model with equivalent lines is used [5]. Each layer is assigned with a part of the radial line. Transmission matrices of equivalent lines and boundaries define electromagnetic waves propagation in each layer and between layers. The method makes unnecessary field calculation inside the Luneburg lens when the far field and radiation patterns are defined. All these essentially reduce computing time.

3. Numerical results

Let us consider radiation properties of the cylindrical Luneburg lens excited by a longitudinal electric dipole. Dipole radiation is uniform in the azimuthal plane if there is no the lens. The lens increases antenna gain and radiation directivity as it is shown in figure 1. Two kinds of structures are under analysis. The first one has the equal thickness of the each layer. Dielectric constant is close enough to equation (1) as follows: ε_1=1.93, ε_2=1.77, ε_3=1.61, ε_4=1.46, ε_5=1.31, ε_6=1.16.

The second structure was optimized for maximum directivity. After optimization permittivity of the layers varies as follows: ε_1=1.88, ε_2=1.72, ε_3=1.56, ε_4=1.40, ε_5=1.24, ε_6=1.08. The radius of the *i*-th layer of the 3λ Luneburg lens is as follows: r_1=0.6λ, r_2=1.35λ, r_3=1.8λ, r_4=2.16λ, r_5=2.49λ, r_6=2.76λ, where λ is the free space wavelength. The optimized lens structure forms more directed radiation. The similar effect is observed in the meridional plane (figure 2). Further this kind of the structure may be used as the first step approximation for Ansys HFSS or CST Microwave studio software.

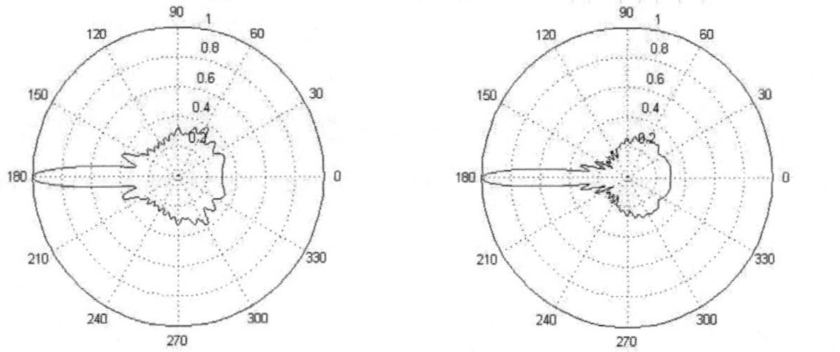

(a) (b)

Figure 1. Radiation pattern in the azimuthal plane of the 3λ cylindrical Luneburg lens excited by the longitudinal electric dipole with equal-thickness (a) and optimized (b) structure.

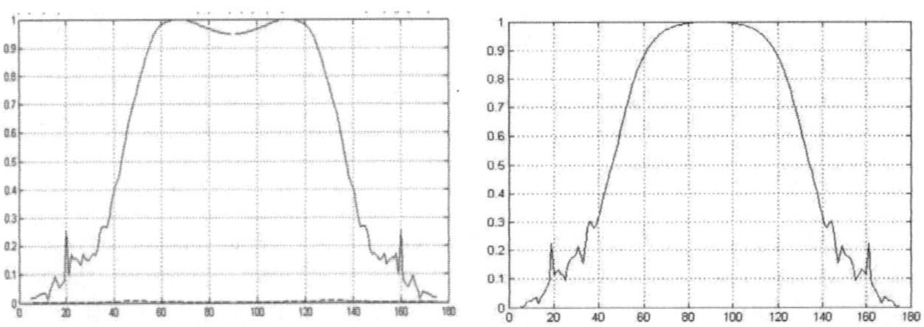

(a) (b)

Figure 2. Radiation pattern in the meridional plane of the 3λ cylindrical Luneburg lens excited by the longitudinal electric dipole with equal-thickness (a) and optimized (b) structure.

Radiation patterns of the 4λ spherical Luneburg lens excited by the aperture antenna with circular polarization are shown in figure 3. Matching between the main polarization and cross-polarized radiation field levels may be simply carried out.

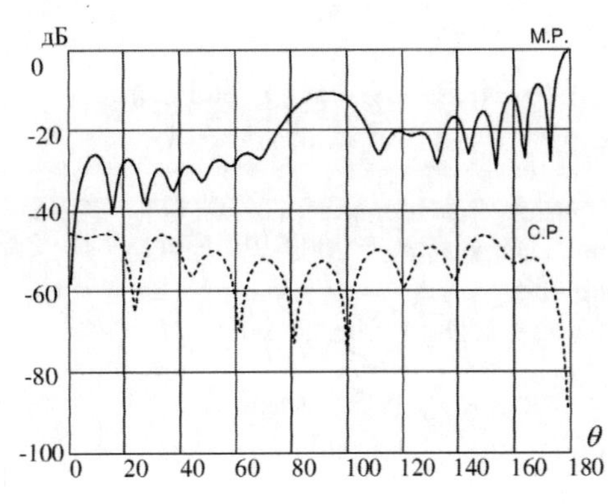

Figure 3. Radiation pattern of the 4λ spherical Luneburg lens exited by the aperture antenna. Main polarization (solid line), Cross-polarization (dash line)

We tested our method in comparison with software Ansys HFSS. The three-layer spherical structure of Luneburg lens was designed. A circular polarized incident wave was excited by a circular horn antenna at the frequency of 25 GHz. The thickness and dielectric constant of the each layer is shown in figure 4. Comparison of radiation patterns at the main polarization calculated by Ansys HFSS and our method are shown in figure 5. The main lobes are very similar. There is difference in sidelobes but it is compensated by a very short processing time (table 1).

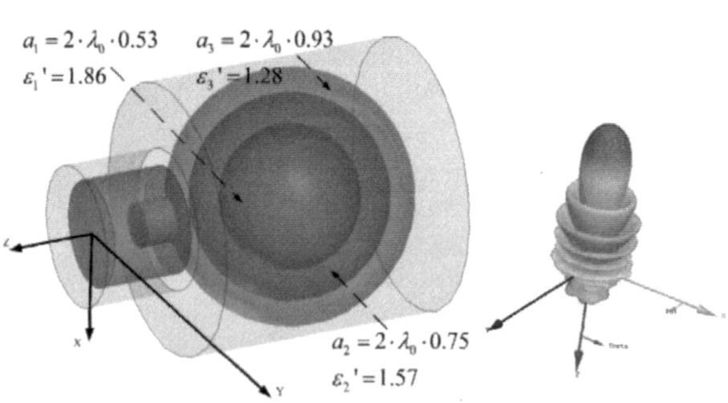

Figure 4. Three-layer spherical Luneburg lens and radiating field in the software Ansys HFSS

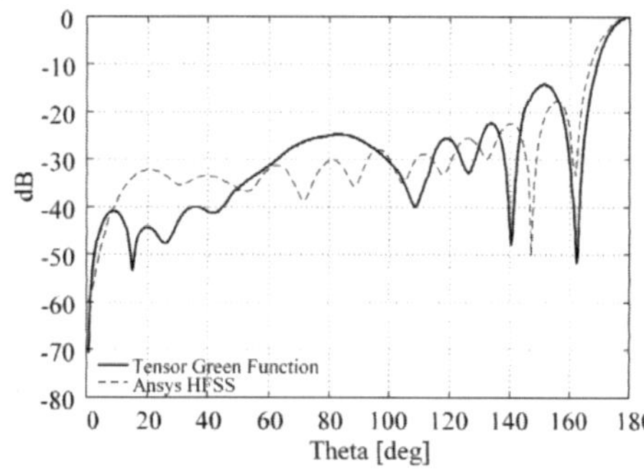

Figure 5. Comparison of radiation patterns of the lens at the main polarization calculated by Ansys HFSS and the Green's functions method

Table 1. Comparison of the beamwidth value and processing time for analysis of 3-layer spherical Luneburg lens made by our method and Ansys HFSS

	$2\Delta\theta^0_{-3\ dB}$		$2\Delta\theta^0_{-10\ dB}$		t, sec	
a	HFSS	TGF	HFSS	TGF	HFSS	TGF
λ	19.2	18.8	33.5	31.4	$2.4\cdot10^3$	1.8
2λ	11.1	10.0	17.5	16.6	$3.7\cdot10^3$	2.3
3λ	6.8	7.0	9.6	11.6	$5.8\cdot10^3$	3.1

The proposed method significantly reduces the computing time for multi-layered lenses analysis in comparison with the most commonly used in antenna design. The first step antenna structure optimization may be performed for a short time. The results may be used as the first approximation for Ansys HFSS, CST Microwave Studio and other software.

4. Conclusion
The efficiency of the universal method based on the Green's functions of layered structures for the spherical and cylindrical Luneburg lenses radiation field calculations is shown.

5. Acknowledgement
This research was supported by grant N 14-19-01396 from the Russian Science Foundation

References
[1] Luneberg R K 1944 *The Mathematical Theory of Optics* (Providence RI Brown Univ. Press)
[2] Baev S R, Gechev S M, Hadjistamov B N and Dankov P I 2008 *Proc. TELFOR, 16th Telecommunication Forum* (Belgrade, Serbia) p 488
[3] Chia T-T and Lim W-Y 2014 *Proc. EuCAP* (Amsterdam, The Netherlands) p 3011
[4] Knyazev S, Kostitsyn Yu and Shabunin S 2011 *Proc. EUCAP* (Rome, Italy) p 498
[5] Panchenko B and Gizatullin M 2007 *Proc. EuCAP* (Edinburgh, UK) p 518
[6] Felsen L and Marcuvitz N 1973 *Radiation and Scattering of Waves* vol 1 (Prentice-Hall, Inc., Englewood Cliffs, New Jersey)

IOP Conf. Series: Materials Science and Engineering **120** (2016) 012012 doi:10.1088/1757-899X/120/1/012012

MITRA 2.0

Vinand Prayag, Nazir Vydelingum, & Girish Kumar Beeharry
Mauritius Radio Telescope, Department of Physics, University of Mauritius, Reduit, Moka
E-mail: vinand28@gmail.com,sn.vydelingum@uom.ac.mu,gkb@uom.ac.mu

Abstract. We give the rationale of building a new innovative radio telescope in Mauritius. The instrument will tackle important astrophysical areas, such as the detection and imaging of the Deuterium line and of recombination lines, as well as bridging a gap in the technological area. The telescope will be composed of low cost dipole antennas, arranged in planar stations with a bandwidth of ~200MHz. The Front-End Electronics will have GaS-HEMT technology Low Noise Amplifiers to constrain the value of the system temperature to a minimum. The signals will be digitized and Fast Fourier Transformed. The post-processing used will be dependent on the science observed. It could be a case of either accumulation or FX correlation.

1. Introduction

The Square Kilometre Array(SKA) radiotelescope will be hosted by South Africa and Australia. South Africa will host the telescope in collaboration with a number of partner countries, mostly in Southern Africa. Mauritius is one of the partner countries. The Multifrequency Interferometry Telescope for Radio Astronomy(MITRA) is a deliberate attempt by some members of the Mauritian radioastronomy community to actively support the SKA consortium by making tangible contributions in science and technological areas. One of its main aims is to study the solar corona, the ionosphere, etc. The MITRA project is designing and developing novel technologies. These are then being shared with a broader global audience. All of this activity will also help in building human capacity around radioastronomy.

A number of SKA related radiotelescopes are being built and commissioned around the world. These are SKA precursors and pathfinders. MeerKAT, ASKAP and LoFAR are a few of them. Others, even without an SKA designation of pathfinder, like the Deuterium Array, are still working on SKA related science and technology. The Southern Hemisphere has been devoid of low and midfrequency radiotelescope for a long time. Presently, the active low frequency instruments are the Murchison Widefield Array(MWA) and Precision Array for Probing the Epoch of Reionization(PAPER). The addition of these telescopes is essential to complement the work done by Northen Hemisphere ones.

The MITRA project will focus on designing and building a sensitive high resolution multi-frequency dual polarity instrument. Multiple stations of low-cost dipoles, such as log-periodic antennas, will be used. The operating bandwidth will be in the range of ~200-~450 MHz. The instrument and each station will be modular in nature. A station will be able to function in its own right. Subsets of the whole instrument could also be used. We are targeting spectral lines such as Deuterium, red-shifted lines of hydrogen, carbon recombination lines and all-sky surveys. These science targets are crucial to the progress of astrophysics. Telescopes such as the Deuterium Array in MIT, Haytstack, USA, and LoFAR, in the Netherlands, are active in some of these areas. Such a telescope in Mauritius will be able to complement, and add a new perspective, to any work done in the North.

Here we present details of our proposed instrument and our intended plan.

2. Radioastronomy between ~200 and ~450 MHz

2.1. Deuterium Line at 327MHz in the Southern Hemisphere

Deuterium is one of the primordial elements formed after the Big Bang. Shklovskii(1952) discussed its astrophysical importance. Subsequently, the line proved to be a determing factor in finding the baryon density of the Early Universe (Wagoner, 1973). Several attempts have been made since then to observe the Deuterium line at 327.4 MHz. Anantharamaiah & Radhakrishnan (1979) and Heiles et al. (1993) placed an upper limit of almost 5.8×10^{-5} for D/H ratio, in absorption, in the Galactic center, and Cassiopeia A. Chengalur et al. (1997) obtained a marginal detection of the emission from the anticenter to place the D/H ratio between 2.9×10^{-5} and 4.9×10^{-5}. Rogers et al. (2005) observed the anticenter for two years and came up with a definitive value of $(2.3 \pm 0.4) \times 10^{-5}$

2.2. Carbon Radio Recombination Lines

Radio Recombination Lines(RRL) are an important probe to study the cold, diffuse, atomic clouds of the Interstellar Medium(ISM). RRLs are particular interesting because dust and contaminants have little impact on them. They are being used to constrain the temperature and density of the cold ISM. Carbon Radio Recombination Lines(CRRLs) have proved to be excellent tracers of the physical conditions in cold, diffuse clouds. Despite its usefulness, only a few surveys have been done on CRRLs at low-frequency(Anantharamaiah and Kantharia 1999, Kantharia & Anantharamaiah 2001; Roshi et al. 2002). More recently Asgekar et al. (2013) used the LOw Frequency ARray (LOFAR) to observe the CRRLs towards Cassiopeia A at very low frequencies. The MITRA telescope with its wide bandwidth will be able to look at CRRL lines with the principal quantum $249<n<288$ which corresponds to a frequency range of $\sim275 < \nu < \sim425$ MHz.

2.3. Red-shifted of the 21cm Hydrogen Line

Neutral Hydrogen emits a radiation at 21 cm, when its ground state goes through a hyperfine splitting. The line occurs when there is a spin-flip transition from the highest state(spins of proton and electron are parallel) to the lowest state(antiparallel spins). With hydrogen being ubiquitous in the universe, it is a wonderful tracer for galaxy surveys. The red-shifted hydrogen line will allow us to map the sky at redshifts of $2.5<z<4$ which corresponds to an Universe being $\sim 1.5 <$ age $< 2.5\times10^9$ years. The early Universe is yet to be fully understood. Surveys of this type are important to study the evolution of the Universe from the epoch of reionization till now.

3. Design Requirements

3.1. Antenna and Front-End Electronics

3.1.1. Antenna A multifrequency and multiscience telescope meant that we had to look at broadband antennas that could offer a relatively linear performance over the band of interest. Log-Periodic Antennas(LPDAs) are the ideal candidates. These antennas respond to a wide range of frequencies, without a significant degradation in the antenna parameters such gain, beam pattern, and SWR. They are relatively easy to construct and are low cost.

Table 1: SUMMARY OF ANTENNA CHARACTERISTICS

Characteristics	Value
Bandwidth	~200MHz-~450MHz
Gain	~9dBi
Polarization	Dual Linear
Collecting Area at 327MHz	$0.6m^2$

3.1.2. Front-End Electronics The Front-End Electronics(FEE) will consist of a matching circuit, a filter and a Low-Noise Amplifier(LNA). The first amplification stage is the most crucial part in constraining the value of the system temperature to a minimum. Each science case will have a dedicated FEE. Since the spectral lines are our primary scientific targets, the bandwith of the filter will be science-limited so that the power noise level of the LNA is at a minimum. Most of the components will be off-the-shelf.

Table 2: SUMMARY OF FEE CHARACTERISTICS

Low Noise Amplifier		Filter	
Parameters	Value	Parameters	Value
Power Consumption	<100mW	Bandwidth	~10MHz
Gain	>20dB	Tolerance	<0.5
Ultra Low Noise Figure	<0.5		

3.2. Array and Station SubArray

The telescope will be composed of 2D sub-stations arranged in a T-shaped array. It will be a 200 m East-West arm with a 400 m South arm. The array will be modular, ie, each station will work independently. Each substation will either be a 4×4 or 5×5 planar array. This arrangement is ideal for our work because it allows for high level of flexibility. The sub-stations can be moved along the array and can either increase or decrease in size depending on the needs. The antennas will be placed on square frames which could also be tilted to scan further on the northern or southern declination. Beamforming can be easily adapted. More than one beam can be obtained from a sub-station.

3.3. Signal Transfer

RFoF photonics technology will be used to carry Radio Frequency(RF) over fibre cables. Outdoor Non Metallic Armoured Singlemode 24 Core fibre cables have already been laid out at the telescope site with a 24 port singlemode Fibre Optic Patch Panel. The RF signal will be modulated by a laser diode and demodulated by a PIN detector. Appropriate amplification stages will be added in case of signal attenuation. The modulators and demodulators will be made using off-the-shelf components. We are expecting to achieve very good results, at a very low-cost.

3.4. Receiver

Each channel will be amplified, filtered, and down converted using a double downconversion superheterodyne receiver. The first and the second one Intermediate Frequency(IF) will be 30 MHz and 10.7 MHz respectively. The IF will then be converted to a digital signal by sampling using a high speed digital acquisition board. The signal will be digitally filtered to a 250 KHz bandwidth in the process. Graphical Processing Units(GPUs) will be used for most of the

Digital Signal Processing(DSP). The digital outputs will be passed to the GPUs for Fast Fourier Transform and correlation. 1024 FFT channels at a frequency resolution of 244 Hz will be obtained.

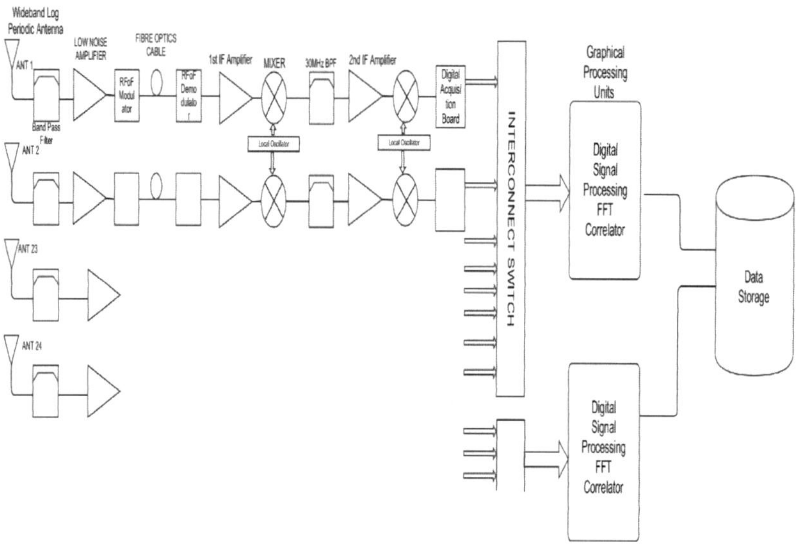

Figure 1. Block diagram of receiver

4. Conclusion

In this paper we have clearly outlined the scientific importance of our work, in the Square Kilometre Array project context. Section 2 goes through the main scientific targets and Section 3 targets the instrumental and technological requirement of the telescope.

[1] I. S. Shklovskii, *Astr. Zh*, 29, 144. USSR, 1952.
[2] R. V. Wagoner, *Ap. J.*,, 179, 343. 1973.
[3] K. Anantharamaiah & V. Radhakrishnan, *A&A*, 79, L9. 1979.
[4] Heiles, C., McCullough, P. R., & Glassgold, A. E., *ApJS*, 89, 271. 1993.
[5] Chengalur, J. N., Braun, R., Burton, W. B., *A&A*, 318, L35. 1997.
[6] Rogers, A. E. E., Dudevoir, K. A., & Bania, T. M., *ApJS*, 133. 1325. 2007.

SwaMURAy – Swapping Memory Unit for Radio Astronomy

Simon Winberg

Department of Electrical Engineering

University of Cape Town

Cape Town, South Africa

simon.winberg@uct.ac.za

Abstract. This paper concerns design and performance testing of an HDL module called SwaMURAy that is a configurable, high-speed data sequencing and flow control module serving as an intermediary between data acquisition and subsequent processing stages. While a FIFO suffices for many applications, our case needed a more elaborate solution to overcome legacy design limitations. The SwaMURAy is designed around a system where a block of sampled data is acquired at a fast rate and is then distributed among multiple processing paths to achieve a desired overall processing rate. This architecture provides an effective design pattern around which various software defined radio (SDR) and radio astronomy applications can be built. This solution was partly in response to legacy design restrictions of the SDR platform we used, a difficulty likely experienced by many developers whereby new sampling peripherals are inhibited by legacy characteristics of an underlying reconfigurable platform. Our SDR platform had a planned lifetime of at least five years as a complete redesign and re-fabrication would be too costly. While the SwaMURAy overcame some performance problems, other problems arose. This paper overviews the SwaMURAy design, performance improvements achieved in an SDR case study, and discusses remaining limitations and workarounds we expect will achieve further improvements.

1. Introduction

A clear trend in radio processing is the increasing demand for higher sampling speeds, greater resolution, greater bandwidth, and more samples to get more from available parts of the EM spectrum. Going for higher speeds, wider bands, and more feeds means bigger data volumes, in turn needing greater processing power and architectural solutions to support application needs. Combined with this tendency is the increasing popularity of reconfigurable computing (RC) platforms [1] as a means to save costs and development time, among other advantages [2]. A development strain can thus develop between increasing application demands and the design constraints, and aging technologies, of the reconfigurable platform around which an application is built. This paper discusses one such situation and the approach we used to work around limitations of our legacy platform during the development of a Software Defined Radio (SDR) application prototype.

The main reason for our decision to follow a RC platforms approach in our prototyping experiment was the time and cost savings it offered. In particular, a newer and more feature-rich analogue-to-digital convertor (ADC) daughterboard offered potential to extend the application scope [3] and to explore new innovation potentials by reusing existing RC platform and code developed in a previous project. This led to throughput challenges that the SwaMURAy module was to address.

2. Background and purpose for development of SwaMURAy

Field programmable gate arrays (FPGAs) have become widely used for SDR and radio astronomy applications. A benefit of FPGAs is their potential for providing highly parallel solutions at a lower cost (for small volumes) in comparison to an ASIC approach [4]. In 2009 we embarked on an in-house design for our own open-hardware RC platform to leverage prototyping benefits of this approach. This led to design of the Reconfigurable Hardware Interface for computatioN and radiO (RHINO) platform.

Development of RHINO started with a thorough needs analysis followed by an optimized selection of parts and tradeoffs made to minimum costs and maximize compatibility with the better known ROACH platform [5]. RHINO is the legacy platform discussed in this paper.

RHINO was designed around similarity to ROACH version 1 to improve opportunity for code reuse and to facilitate transition from training contexts and small scale and low budget prototyping experiments to larger scale endeavors using ROACH. There are three versions of RHINO, namely RHINO versions and 1.01, 1.02 and the Zynq-based RHINO-SDR. The platform related to this paper is RHINO 1.02. An overview of RHINO platform is shown in Figure 1.

Like ROACH 1, the RHINO architectural design is separated into three subsystems: the control processor, the FPGA subsystem, and the monitoring and management subsystem. In the case of RHINO the control processor is an ARM Cortex microcontroller instead of a PowerPC. The RHINO has a Spartan 6 FPGA, the performance and capacity of which is less than the ROACH's Vertex 5. This Spartan was chosen because, at the time of designing the board in 2010, it was determined to be the lowest cost and best performing option using tools that our design team and our industry partners were familiar with [6].

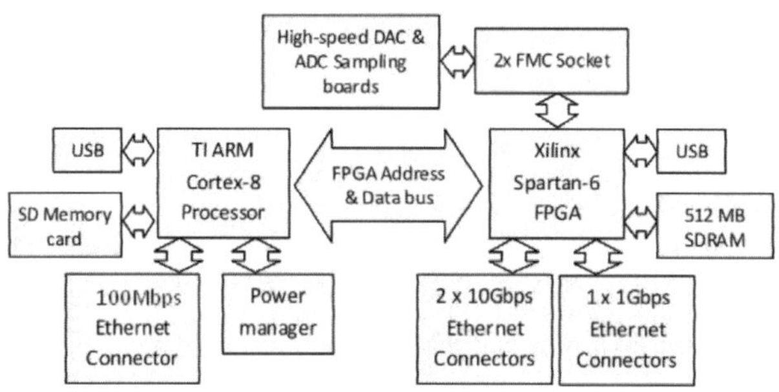

Figure 1: Overview of the RHINO platform.

The RHINO has an analogous FMC daughterboard plug-in design as found on ROACH. The monitoring and management subsystem of RHINO is dissimilar to the ROACH. As is the case with earlier versions of ROACH, the BORPH operating system is also used on RHINO [7,8].

The root cause for the SwaMURAy occurred during an attempt to prototype an experimental system that would simultaneously run two semi-independent operations on a single RHINO platform. In order to run this application we planned to use of a commercial-off-the-shelf, high-speed sampling board, namely the FMC150 from 4DSP. Previously we had focused on narrower band acquisition and more RF operations happening in hardware, for which sub-10MS/s sampling rates were adequate. The FMC150 can sample two channels, up to 250MS/s. The intended application involved wideband sampling of multiple FM channels (on port A) and sensing solar events on a finely tuned separate channel (port B). Digital down converters (DDCs) were to be used to cut down the data volume and extract individual FM channels sent on to the host PC. The main problem was that the existing processing blocks could not process the data quickly enough; hence a solution for this was needed.

3. Methodology

The major design requirements to be achieved using the SwaMURAy module were: a) sampling at 100MS/s with 14 bit samples, and b) to allow the data to be processed by digital filters achieving at least 20Mb/s post filtering output speeds, i.e. the speed to send partially processed data to the host PC.

The SwaMURAy module was developed by first following a simulation-based approach whereby we developed a simulated ADC interface module which read test signals from a file and mimicked interfacing characteristics of the actual ADC peripheral. Similarly a mimicked Ethernet module was developed to stream processing results to a binary file. Once a working system had been developed and tested in simulation, the process of porting to the target hardware was commenced. This was a lengthy process of revisiting, improving and retesting the simulation and target implementations.

4. Design of SwaMURAy

The SwaMURAy is designed around being a portable gateware module useful as a scheduling and intermediary flow-control subsystem used to split data, sampled at high stream, into two or more separate data blocks or pipelined processing channels. The module has configuration options, set in the code: it can either write to memory, to dual-buffered RAM (DRAM) or to BRAM; or it can send data via a parallel connection to processing modules. The first case is more complex as integration with the external memory banks needs to be done and necessary HDL code needs to be provided for controlling these operations (i.e., using the core generator in the ISE and wiring up the links). The SwaMURAy module is dependent on a 'gather' module to which its input lines are wired. The gather module provides an interface wrapper to an ADC driver or another data source that the SwaMURAy will then split into separate output blocks or streams. A block diagram of the SwaMURAy is shown in Figure 2.

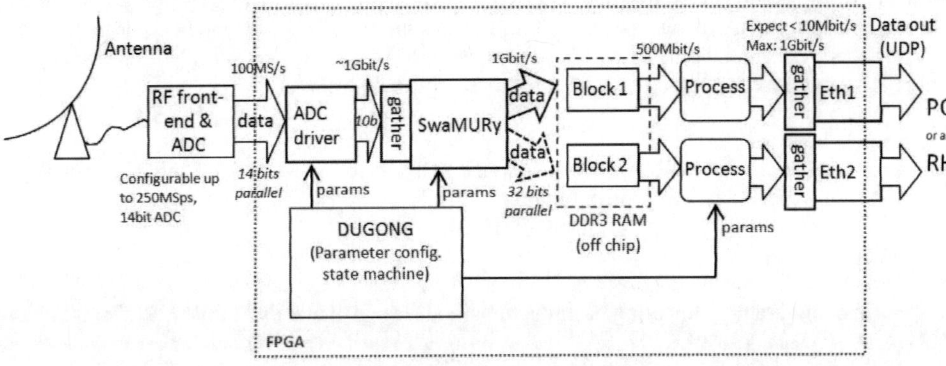

Figure 2: Structure of an FPGA-based system using the SwaMURAy.

The DUGONG module [9], bottom of Figure 2, is an optional HDL state machine to configure SwaMURAy. In this application it can set block size at run time to specify when to swap from sending to one block/stream to another. This setting can alternatively be hard-coded in the SwaMURAy code.

5. Results and conclusions

Initial simulation-based testing was performed using iVerilog, which was limited to functional simulation. A test bench was coded to evaluate SwaMURAy using pre-generated sinusoidal signals and streamed via file to substitute for the ADC input. The SwaMURAy swapped between sending data to two DDC processing modules. The simulation results worked correctly, showing no glitches in the processing caused by potential clocking, synchronization or handshaking mistakes.

Moving to the hardware implementation was more involved. The main challenge was integrating the DRAM control logic. A drawback of the DRAM controller for the RHINO is that it utilizes an intermediary cache, a BRAM buffer, and a state machine controller that accepts a starting address and block transfer size to send/receive data to DRAM. The RHINO has two independently addressable memory banks for which the read/write transfers needs to be carefully scheduled. A maximum DRAM cache size of 1024 samples (16-bit words for each) was used. Sending data from the RHINO to the PC, and to validate the processing, an HDL Ethernet module was incorporated since the RHINO has no onboard Ethernet peripheral. The OpenCores Ethernet 1Gb/s MAC was utilized to this end.

From the testing, it was found that the desired 20Mb/s output via Ethernet from the RHINO to a PC was achieve. However only output from the DDC results of the wideband FM sampled channels were streamed to the PC – the H1 sampled data was not streamed simultaneously but instead had to be multiplexed through the single Ethernet port if the data was required. The FFT for H1 processing was not done in gateware. In performance analysis of this sampling chain of ADC to SwaMURAy to BRAM input of DDC (two DDCs were swapped between) and out to the Ethernet and then to the PC, the maximum sustainable input ADC sampling rate was 61.4MSps, corresponding to a raw data transfer speed of 982Mbps (14 bit sample words padded to 16 bits). Note that this was bypassing the DRAM and having the SwaMURAy write directly to BRAM of the DDC awaiting input, and there was no FFT stage after this on the FPGA; the FFT was done on the host PC. Accordingly, the desired ADC speed of 100MSps with the whole processing chain operating on the FPGA was not fully achieved. Further optimization is needed to get closer to the desired performance level. Furthermore, the SwaMURAy DRAM mode could only operate reliably at slower speeds as data was lost during switchover periods. Nevertheless, simultaneous processing of sampled data ran successfully with SwaMURAy operating as planned for processing the FM channels. In simulation the 100MS/s case with simultaneous H1 processing was successful.

Plans for further work focus at getting the higher sampling rates to work reliably, for which a major problem was synchronizing of data sent from the SwaMURAy to DRAM whereby new data could start overwriting previous memory that had not been fully processed. Any processing module having to re-access data in DRAM was not possible in this design; the FIFO read index could not be reset to allow a second pass through the data; this feature needs to be added. Accordingly, improvement to the SwaMURAy interface is needed to account for these desirable features.

6. Acknowledgement

Thanks to members of the SDRG for input on RHINO, and to the MeerKAT and National Research Fund (NRF) of South Africa for project funding.

References

[1] Todman T J, Constantinides G A, Wilton S J E, Mencer O, Luk W and Cheung P Y K 2005 Reconfigurable computing: architectures and design methods *IEE Proceedings-Computers Digit. Tech.* **152** 193–207

[2] Hauck S and DeHon A 2010 *Reconfigurable computing: the theory and practice of FPGA-based computation* (Morgan Kaufmann)

[3] Kuo S M, Lee B H and Tian W 2013 *Real-Time Digital Signal Processing: Fundamentals, Implementations and Applications* (John Wiley & Sons)

[4] Lin M-B 2008 *Digital System Designs and Practices: Using Verilog HDL and FPGAs* (Wiley Publishing)

[5] Scott S 2011 *RHINO: Reconfigurable Hardware Interface for Computation and Radio* (University of Cape Town)

[6] Winberg S, Langman A and Scott S 2011 The RHINO platform: charging towards innovation and skills development in software defined radio *South African Institute for Computer Scientists and Information Technologists* (Cape Town: ACM) pp 334–7

[7] Hamilton B K and So H K H 2010 BORPH: Operating system support on the NetFPGA platform *North American NetFPGA Developers Workshop*

[8] Hamilton B and Langman A 2011 *BORPH port for RHINO*

[9] Bridges M 2015 *DUGONG* (University of Cape Town)

IEEE Radio and Antenna Days of the Indian Ocean (IEEE RADIO 2015) IOP Publishing
IOP Conf. Series: Materials Science and Engineering **120** (2016) 012014 doi:10.1088/1757-899X/120/1/012014

Low Frequency Radio Experiment (LORE)

P. K. Manoharan[1,2], Arun Naidu[2], B. C. Joshi[2], Jayashree Roy[2], G. Kate[2], Kaiwalya Pethe[3], Shridhar Galande[3], Sachin Jamadar[3], S. P. Mahajan[3], R. A. Patil[3]

[1] Radio Astronomy Centre, National Centre for Radio Astrophysics (TIFR), Ooty, India
[2] National Centre for Radio Astrophysics (TIFR), Pune, India
[3] Department Of Electronics and Tele-Communication, College of Engineering, Pune, India

E-mail: mano@ncra.tifr.res.in

Abstract. In this paper, we present a case study of Low Frequency Radio Experiment (LORE) payload to probe the corona and the solar disturbances at solar offsets greater than 2 solar radii, i.e., at frequencies below 30 MHz. The LORE can be complimentary to the planned Indian solar mission, "Aditya-L1" and its other payloads as well as synergistic to ground-based interplanetary scintillation (IPS) observations, which are routinely carried out by the Ooty Radio Telescope. We discuss the baseline design and technical details of the proposed LORE and its particular suitability for providing measurements on the detailed time and frequency structure of fast drifting type-III and slow drifting type-II radio bursts with unprecedented time and frequency resolutions. We also brief the gonio-polarimetry, which is possible with better-designed antennas and state-of-the-art electronics, employing FPGAs and an intelligent data management system. These would enable us to make a wide range of studies, such as nonlinear plasma processes in the Sun-Earth distance, in-situ radio emission from coronal mass ejections (CMEs), interplanetary CME driven shocks, nature of ICMEs driving decelerating IP shocks and space weather effects of solar wind interaction regions.

1. Introduction

Number of outstanding problems in solar physics demand a much deeper understanding and require the quantitative knowledge of the structure and evolution of active regions, flares, filaments, coronal mass ejections (CMEs), and formation and flow of solar wind at heights above the photosphere and their consequences in the heliosphere. Such studies are important for an improved understanding of solar eruptions and the 'Sun-Earth' connection events, i.e., space weather effects at the Earth or elsewhere. The proposed Aditya-L1, the first Indian space solar coronagraph, aims to provide unique observables of direct relevance to the understanding of physical processes at the near-Sun region below 2 solar radii. The low frequency radio measurements ($\sim$20 kHz to below 30 MHz) at farther heights are to study the complex relationship between CMEs, prominences, interplanetary shocks, radio emissions (fast drifting type III and slow drifting type II radio bursts), and associated nonlinear processes in the inner heliosphere. Such measurements as a function of time or distance from the Sun and their corresponding CME track observed with the Ooty IPS will be useful to understand the shock acceleration of particles in front of the CME.

2. Scientific Objectives

A high-sensitivity low-frequency radio instrument (covering ~20 kHz to 30 MHz) is essentially required for the understanding of complex relationship between various solar and interplanetary phenomena, as well as their related micro-processes. The lower part of this waveband (below 20 MHz) is reflected by the ionosphere and hence cannot be observed from Earth. Most of the low-frequency radio emission is due to coherent plasma processes and it is closely related to the plasma frequency/density of the solar wind through which the electron beam propagates. Since the plasma density decreases away from the Sun, the plasma frequency also decreases (as $\sqrt{N_e}$), hence radio emission at longer wavelength originates at greater distances from the Sun. At the above frequency range, a rich variety of radio bursts of solar origin can be observed that trace particles produced by energetic phenomena through the solar corona and in the interplanetary medium. Additionally, observations of kilometric waves are valuable to infer processes, such as (i) generation of radio waves, (ii) cascade of plasma turbulence, (iii) reconnection at flux-tube boundaries, (iv) acceleration of particles, etc.

The LORE aims at a high sensitivity radio instrument to address the issue of using radio signals as the potential probe to study the onset and lift-off of CMEs and associated solar processes, such as prominences, shocks, electron acceleration (variety of radio bursts). Some of the key science objectives of the LORE payload are:

(i) *Nonlinear plasma processes*: LORE can address the nonlinear plasma processes in beamed plasma and collisionless shock propagation in the solar wind, and their link to the excitation mechanism of radio emissions. It will allow the high time/frequency resolution observations of in-situ data in the vicinity of CME driven shocks and in the source regions of type III radio bursts along with the Ooty IPS measurements.

(ii) *CME in-situ radio emission*: The detection of weak radio spectral emission near the local plasma frequency (i.e., plasma line) by LORE can be extremely useful to obtain information on electron density and temperature, particularly when a CME passes through the LORE antennas. It will also enhance our understanding on the relationship between waves generated within the CME and in its front.

(iii) *Remote probe of Interplanetary CME-driven phenomena*: Shocks produced by CMEs are efficient accelerators of energetic particles especially close to the Sun. However, their impact at 1 AU is still not negligible and can compress the Earth's magnetosphere moving the radiation belts to lower altitudes. Despite decades of observations on these critical heliospheric structures, much is still unknown. For example, the study by Manoharan et al. [1] combined the radio and interplanetary measurements and provided the first time report of tracking of the Bastille Day event all the way from the atmosphere of the Sun to 1 AU [1,2]. However, it lacked in giving details of structures within the ICME. The LORE observations would contribute to the significant advances in this field. The sensitive low-frequency radio spectrum will be useful to track the shock produced by the CME and to link the flux of solar energetic particle acceleration site along the "Parkar (Archimedean) spiral" of the interplanetary magnetic field.

As a complementary payload on Aditya-L1 mission, the low-frequency solar radio spectral instrument, LORE, is not only limited to the above scientific objectives, but can also serve in various studies to understand the physical processes of the solar transients and their associated interplanetary phenomena.

3. Baseline Design

As mentioned in the previous section, the primary aim is to obtain high temporal and frequency resolutions of the radio emission generated in association with the type II and III solar radio bursts below 30 MHz. Observations will be accomplished using three non-planar mutually orthogonal monopole antennas mounted at an angle of 45° to the spacecraft

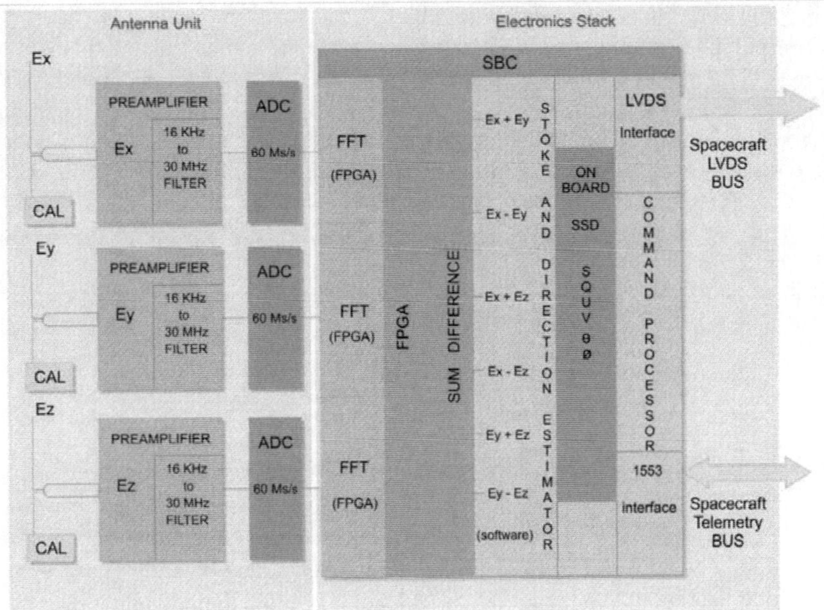

Figure 1. Block diagram of the required signal processing and electronics systems

either in the Sun-ward or in the anti-Sun-ward direction. A combination of measurements from the three-antenna system in a three-axis stabilized spacecraft has been shown to provide measurements of four Stokes parameters and two direction coordinates of the burst [3]. The three-antenna configuration has been chosen for implementation of direction finding along with the polarimetery of the burst emission.

For the given frequency range of LORE, the monopoles proposed to be employed for this purpose are short dipoles of 7-m length. A scaled version of the antenna operating at 70-MHz has been developed by us recently (refer to Pethe et. al., this issue) and computation of electromagnetic simulations are currently in progress for the final prototype. An attempt to optimize this antenna configuration with detailed electromagnetic simulation is being carried out to obtain the uniform sensitivity and gain over the entire 30 MHz bandpass. It may be noted that previous missions lacked such a uniformity across the bandpass. Each antenna will be fabricated from heat treated tightly rolled Be-Cu sheets.

One of the important challenging aspects of the radio astronomy is the sampling of large bandwidth and processing on short time scales. The current technology allows direct digitization of entire band with 60 Msamples/sec sampling and synthesis of channels with required time and frequency resolutions with digital signal processing. This will allow the experiment to have much superior time and spectral resolutions than the previous missions allowing high quality observations motivated by our science goals. We also propose an electronics design using new technologies to provide greater flexibility and larger number of observing modes than the previous experiments. Such routine sampling is already employed in the Ooty Radio Telescope and the GMRT for pulsar studies [4]. The required signal processing has already been implemented for the above ground-based telescopes and is being suitably modified for the signal processing required for our payload and be ported on space qualified hardware. The conceptual design for the electronics, described below, is based on this experience. A block diagram of the required signal processing and electronics is shown in Figure 1.

Each antenna will be backed by a high gain pre-amplifier, which will be placed close to the antenna unit. The signal from each pre-amplifier will be directly digitized at a rate of 60 Msamples/s and processed in the electronics unit. This was not attempted in previous missions due to unavailability of high speed digitizers in the last decade and will be one of the significant differences in our design. It is proposed to use a design employing FPGAs for our signal processing. The FFT of the sampled data, followed by an appropriate accumulation, will be carried out to obtain desired frequency resolution for the different modes of operation as discussed below.

The signals from the three-antenna system will be combined and an algorithm encoded in a space qualified single board computer (SBC) will be applied to obtain the required gonio-polarimetery. Such SBCs along with included FPGA are being evaluated by us right now from the point of view of required signal processing capabilities. The aim of this design is to do considerable signal processing on board, thereby reducing the volume of data to be down-linked. It is to be noted that powerful SBCs for embedded computing were not available, when previous similar missions, such as WIND/Waves and STEREO/Waves, were designed (in late 1980s and 1990s respectively) with a limited technology available then. A larger on-board processing will allow us a higher cadence (time/frequency resolution) than these instruments, enabling a much higher science output. The use of FPGA also allows for an improvement in calibration and processing by limited one-time reload of programs for very specific science goals. The SBC will also act as a supervisor for the electronic stack by responding to ground based commands. Greater intelligence on-board will also allow autonomous event triggered mode change, eliminating the need for ground based commands, for important solar events. The algorithms for this are being developed after analysis of the data on similar event from previous missions.

There are two broad modes of observations proposed for the LORE. During energetic solar events, such as solar flares, CMEs, etc., type II and III bursts will be monitored with higher time and frequency resolutions, generating higher data rates and volumes for short durations. This mode of observation is called high rate science (HRS). The 30-MHz bandpass will be subdivided digitally into 2048 channels, providing a frequency resolution of $\sim$15 kHz. Accumulation of about 1500 such spectra will allow 100 ms temporal resolution for each channel. On-board processing will provide four stokes parameters and two direction angles with an 8-bit resolution giving a peak data rate of 1 Mbps and a total volume of 150 Mbytes. On all occasions other than during energetic solar events, the data will be acquired with 512 channels (about 60 kHz frequency resolution) and once every 500 ms at a data rate of 6 kbps, giving a total volume of 4 Gbytes per day. The data rate is much lower for this low rate science (LRS) mode, but overall data volume is large. We are examining strategies to decide a more useful subset of these data so that the data volume would be further reduced. The interface to the electronics for telemetry will be using a MIL-STD-1553 bus interface.

The antenna configuration and pre-amplifiers have been developed as of now and the prototyping the digital electronics is currently in progress.

Acknowledgments

This work is supported by XII plan grant TIFR-12P0716

References

[1] Manoharan P K, Tokumaru M, Pick M, Subramanian P, Ipavich F M, Schenk K, Kaiser M L, Lepping R P, and Vourlidas A 2001 *Astrophysical Journal*, **559** 1180
[2] Manoharan P K 2010 *Solar Physics* **265** 137
[3] Cecconi B and Zarka P 2005 *Radio Science* bf 40 RS3003
[4] Naidu A, Joshi B C, Manoharan P K, and Krishnakumar M A 2015 *Experimental Astronomy* **39** 319

IOP Conf. Series: Materials Science and Engineering **120** (2016) 012015 doi:10.1088/1757-899X/120/1/012015

Statistical Performances of Resistive Active Power Splitter

Sébastien Lalléchère, Blaise Ravelo, Atul Thakur

Université Clermont Auvergne, Université Blaise Pascal, Institut Pascal, 4 avenue Blaise Pascal, BP 80026, 63178 Aubière Cedex, France

E-mail: sebastien.lallechere@univ-bpclermont.fr

Abstract. In this paper, the synthesis and sensitivity analysis of an active power splitter (PWS) is proposed. It is based on the active cell composed of a Field Effect Transistor in cascade with shunted resistor at the input and the output (resistive amplifier topology). The PWS uncertainty versus resistance tolerances is suggested by using stochastic method. Furthermore, with the proposed topology, we can control easily the device gain while varying a resistance. This provides useful tool to analyse the statistical sensitivity of the system in uncertain environment.

1. Introduction

The power splitter (PWS) is one of key microwave devices that can be found in different stage radio-frequency communication system. The PWS is usually implemented in three ports: the two output ports are in same magnitude, 0° phase between any two output signals. Until now, many topologies of PWS have been already proposed in the literature and the majority is in passive circuit [1]. So, significant losses can appear, thus this proposition of active PWS topology which enables to compensate these losses and even to amplify the signal. Recently in [2], a new structure of the broadband PWS based on the topology of the active cell has been introduced [3]. Signal distortion occurs due to the gain variation, and limits usable frequency band. Consequently, a cell constituted by FET in cascade with a shunt resistor is suggested in this paper which is organized as follows; section 2 is devoted to the theoretical study of the proposed PWS. The simulation results are explored in section 3 jointly with a sensitivity analysis over resistive input parameters, before conclusion and prospects.

2. Theory of the proposed PWS

First and foremost, the PWS architecture is composed of the black box of circuits C_1 and C_2, and the input matching shunts resistor R_m. To synthesize this balanced active PWS, the circuits C_1 and C_2 should be identical (Fig. 1). Similarly to the approach introduced in [4], the elementary active cells should be comprised of a FET loaded by a shunt resistance named FET-R network. This cell is particularly advantageous for the PWS interbranch isolation. It is noteworthy that, for the sake of simplification along the paper, the considered FET is comprised of voltage controlled current source with a transconductance g_m, in cascade with the drain-source resistor, R_{ds}. The synthesis formulas for the case of single- and two-cell(s) in cascade in function of the desired gain and matching levels are expressed in the following.

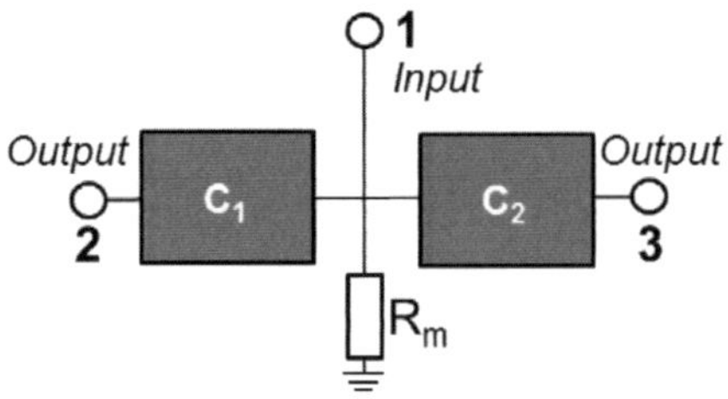

Figure 1. Architecture of the proposed PWS.

By taking into account the matching resistor R_m, we can establish the single cell two-ports S-parameters. The resistance values can be determined with the following synthesis relations:

$$R_m = Z_0 \left(1 + S_{11}\right) / \left(1 - S_{11}\right), \tag{1}$$

$$R = Z_0 R_{ds} S_{21} / \left[g_m Z_0 R_{ds} \left(1 + S_{11}\right) - S_{21} \left(Z_0 + R_{ds}\right)\right]. \tag{2}$$

We can synthesize the resistor value R knowing S_{22}:

$$R = Z_0 R_{ds} \left(1 - S_{22}\right) / \left[S_{22} \left(Z_0 + R_{ds}\right) - Z_0 + R_{ds}\right], \tag{3}$$

and in this case, the transmission loss S_{21} must be equal to: $S_{21} = -g_m Z_0 \left(1 + S_{11}\right) / \left(1 - S_{22}\right)$.

In the case of two-stages cascaded elements, the two cells can be different or identical. For the different FET-R network, it can be established that the transmission gain is expressed as:

$$S_{21} = \frac{2 g_m^2 R_{ds}^2 Z_0 R_m R_1 R_2}{\left(R_m + Z_0\right)\left(R_1 + R_{ds}\right)\left(Z_0 R_2 + Z_0 R_{ds} + R_2 R_{ds}\right)}. \tag{4}$$

Therefore, by using (1) and (3) to determine R_m and R_2, the intermediate resistance can be synthesized from the desired input/output matching levels and gain: $R = 2R_{ds}S_{21}/\left[2S_{21} + g_m^2 Z_0 R_{ds}\left(1 + S_{11}\right)\left(1 - S_{22}\right)\right]$. In such a case, the maximum gain can be expressed in function of the FET parameters g_m and R_{ds}: $S_{21max} = g_m^2 Z_0 R_{ds}\left(1 + S_{11}\right)\left(1 - S_{22}\right)/2$.

Under the assumption $R_1 = R_2 = R$, the synthesis relation can be expressed as follows knowing S_{11} and S_{21}:

$$R = \frac{2R_{ds}\left[S_{21}\left(R_{ds} + 2Z_0\right) + R_{ds}\sqrt{S_{21}\left(S_{21} + 4g_m^2 Z_0^2\left(1 + S_{11}\right)\right)}\right]}{2\left[Z_0 g_m^2 R_{ds}^2\left(1 + S_{11}\right) - S_{21}\left(R_{ds} + Z_0\right)\right]}. \tag{5}$$

3. Numerical investigations

3.1. Ideal proof-of-concept

As a proof-of-concept (POC), the detailed circuit schematic depicted in Fig. 2a was designed and simulated using SPICE® environment.

The S-parameters of the commercial FET $CF004 - 01$ are used during the numerical computations. The active PWS simulation results show a rather constant 0° output phase (0° ± 5°), insertion losses above −1.5dB and an excellent isolation below −30dB for all three ports, for a bandwidth from 1GHz to 3GHz (see Fig. 2b). The system acts as an active PWS using in each arm a two-stages active FET-R cell. It can be pointed out that the outputs of each branch are in phase to the output in the ideal case. Thanks to synthesis relation (4), the insertion loss can change inversely with either R_m, R_1 or R_2. Since insertion loss S_{12} of the FET $CF004 - 01$ is relatively low, R_1 can be varied to fit S_{21} without changing the return losses. The parametric analysis based on the swept variation of the matching resistance R_1 from 40Ω to 55Ω was addressed (not shown here). It can be emphasized that the insertion

IOP Conf. Series: Materials Science and Engineering **120** (2016) 012015 doi:10.1088/1757-899X/120/1/012015

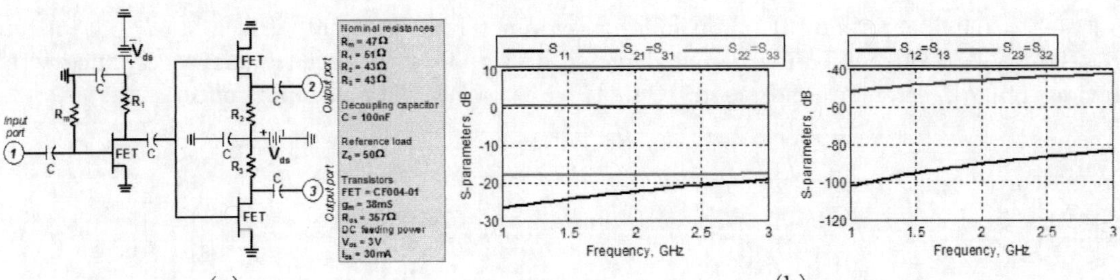

(a) (b)

Figure 2. (a) Schematic of the active PWS POC with nominal values. (b) S-parameters given from the active PWS POC.

losses $\Delta S_{21} = \Delta S_{31} = f(R_1)$ presented about 2.5dB maximal variation. Then, the return losses $\Delta S_{11} = \Delta S_{22} = \Delta S_{33} = g(R_1)$ are rather less sensitive with 0.5dB maximal variation. Although parametric study is useful, the assessment of active PWS performances may be improved throughout statistical investigations in the following.

3.2. Statistical and sensitivity analyses

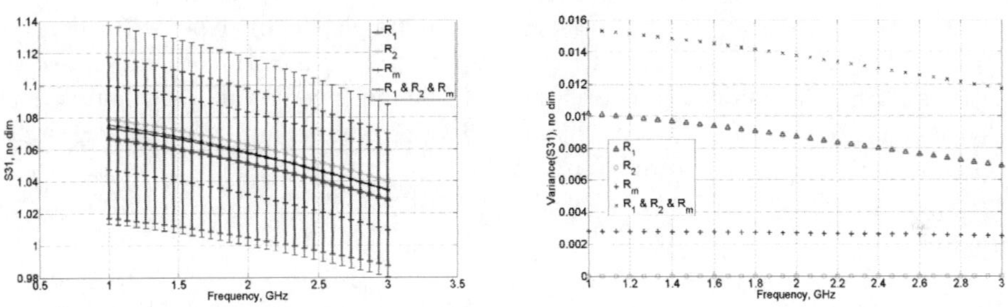

(a) (b)

Figure 3. (a) Mean trend and CIs (error bars) for S_{31} coefficients. (b) Influence (e.g. regarding S_{31} variance) of each random parameter (aka R_1, R_2, R_m in Fig. 2) in comparison to whole model (i.e. including previous RVs)

A sensitivity analysis is proposed following previous numerical model and considering 4 different cases including parameters considered as random one at a time (i.e. R_1, R_2 and R_m respectively), and entirely randomly chosen (Gaussian distribution with mean 51Ω and standard deviation 5.1Ω). Among the diversity of stochastic techniques available [5, 6], crude Monte Carlo (MC) was used. Fig. 3a depicts the statistical results obtained under Gaussian assumption and well-known Student's t-distribution as follows:

$$CI_{\alpha}^{\mu} = \left[\bar{x} - t_{1-\alpha}^{n-1} \sqrt{\frac{var}{n}}, \bar{x} + t_{1-\alpha}^{n-1} \sqrt{\frac{var}{n}} \right], \tag{6}$$

where CI_{α}^{μ} stands for confidence interval (CI) of mean μ from population (here S_{31} parameter), $t_{1-\alpha}^{n-1}$ represents $(1 - \alpha)$-quantile of Student's distribution with $n - 1$ degrees of freedom, $\bar{x}$ and var are mean and variance computed from n samples. Fig. 3 is obtained from 25 MC simulations regarding 98% testing confidence interval with Student's distribution (i.e. $\alpha = 0.01$

and $1 - \alpha = 0.99$ in relation (6)). Transmission parameter S_{31} is mainly driven by R_1 parameter over the whole frequency bandwidth as depicted in Fig. 3b with variances. In accordance with previous remark, errors around mean trends (Fig. 3a) allow ranking inputs effects.

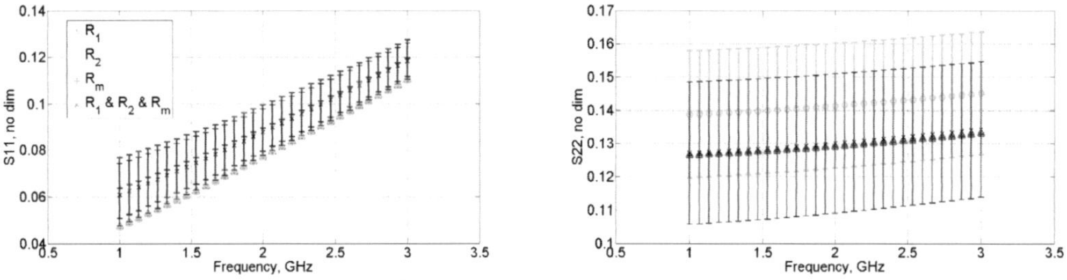

(a) (b)

Figure 4. Reflection coefficients S_{11} (a) and S_{22} (b) from random models including: R_1 (red), R_2 (green), R_m (blue) parameters solely, and together (black). Error bars are given to assess dispersion around mean trend following Student's distribution.

Based upon relation (6), Fig. 4 provides CI around mean S-parameters trends (x/y-linear drawing). Reflection coefficients are not influenced by R_1 parameter whereas R_m and R_2 plays a key role respectively considering S_{11} and S_{22} (which may be physically expected). This overview of system's sensitivity clearly demonstrates the importance of modeling random variations of inputs (e.g by ranking effects of parameters) since this may involve different drifts: from -0.2 to 1.2dB S_{21}-mean, S_{32} isolation is lower than -41dB (not shown here), and mean trends from reflection parameters S_{11}/S_{22} are lower than -15dB.

4. Conclusion

An active resistive PWS is proposed. The analytical relation indicating how to synthesize the PWS resistive parameters are established. The POC based on two-stages circuit is analyzed. The sensitivity of PWS S-parameters in function of the resistive data are analyzed. Further works relying on stochastic methods [5] should efficiently provide accurate results to ensure security levels when computing S-parameters, leading to enforce the reliability of systems [6].

References

[1] Ung-Hee Park P and Jong-Sik Lim L 2005 A 700- to 2500-MHz microstrip Balun using a Wilkinson divider and 3-dB quadrature couplers, *Microwave and Optical Technology Letters*, 47(4) 333–335
[2] Ravelo B 2012 Synthesis of N-way active topology for wide-band RF/microwave applications, *Int. J. Electronics*, 99(5) 597–608
[3] Bharj S, Tan S and Thompson B 1989 A 2-18 GHz 180 degree phase splitter network, *in Proc. IEEE MTT-Symp. IMS*, 952–959
[4] Lu L, Liao Y and Wu C-R 2005 A miniaturized Wilkinson power divider with CMOS active inductors, *IEEE MWCL*, 15(11) 775–777
[5] Cheng X and Monebhurrun V 2014 Uncertainty evaluation in numerical modeling of complex devices, *IOP Conf. Series: Materials Science and Engineering*, 67 1–4
[6] Larbi M, Besnier P and Pecqueux B 2015 Probability of EMC failure and sensitivity analysis with regard to uncertain variables by reliability methods, *IEEE Transactions on EMC*, 57(2) 274-282

IOP Conf. Series: Materials Science and Engineering **120** (2016) 012016 doi:10.1088/1757-899X/120/1/012016

Modelling of Polarization Rotator of the Broadband Scanning Antenna Array

A I Semenikhin[1], D V Semenikhina[1] and Y V Yukhanov[1]

[1]Southern Federal University, Taganrog, Russia

E-mail: d_semenikhina@mail.ru

Abstract. In the paper on the basis of the finite-element method an operation of broadband polarizer in an aperture of a scanning antenna array is analyzed in the frequency range 8 to 18 GHz. The effect of broadband four-grid 45° rotator of polarization on the characteristics of the linear array of parallel-plate waveguides (VSWR, gain, radiation pattern) in scan sector ±45° in E-plane is evaluated.

1. Introduction

In antenna techniques various microwave polarizers of electromagnetic waves [1] are widely used. Antenna arrays with oblique polarization are applied in modern devices of radio technical monitoring of RF radiation sources. Broadband scanning arrays and multipath antenna arrays with oblique polarization can be realized on the basis of radiators with horizontal or vertical polarization. A turn of a polarization plane in subsequent can be carried out by means of polarization rotators (polarizers) located in an aperture of antenna array.

The multilayer UWB transmission polarizers [2-4] are most studied. They implement a rotation of the polarization of the electromagnetic wave transmitted over a wide frequency range and a large sector of angles of incidence. The designing of polarizers can be executed by means of rigorous methods [2], the CAD programs [5] and an approximate impedance method [6].

Properties of broadband polarizers on the basis of multistage dense wire grids with the changing slope angle of conductors from a grid to a grid are widely investigated for a case of normal incidence of plane wave on a polarizer.

The aim of this work is to analyze of operation of antenna array with the polarizer in the given sector of scanning of array and to find out an effect of a polarizer on characteristics of antenna array (in the form of the linear waveguide array) in this sector of angles.

2. The model of a polarization rotator

Let us analyze the polarizer operation in the given sector of scanning in frequency range 8 to 18 GHz. The model of the 45° polarization rotator has four grids (1 mm grid step) and three separator layers ($\varepsilon = 1.02$). The thicknesses of separator layers ($L1$, $L2$, and $L3$) and grids ($G1$, $G2$, $G3$, and $G4$) were variable parameters. The left part of figure 1 and table 1 show the design and dimensions of the polarizer as result of the optimization. The model is analized by means of the finite-element method (FEM) using HFSS. In a case of normal incidence the characteristics of a polarizer were calculated by an impedance method [6].

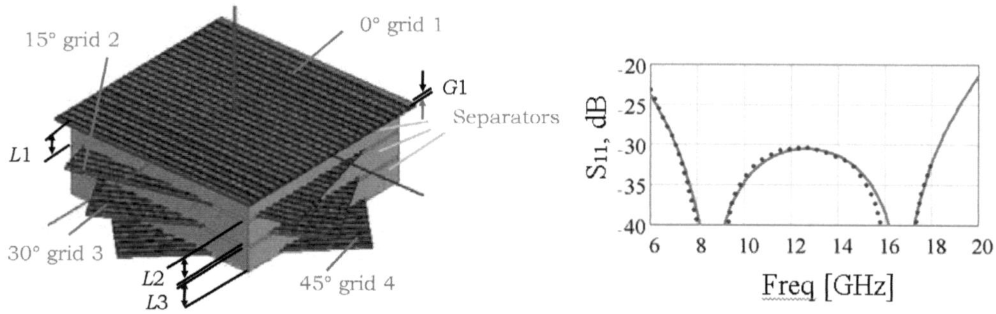

Figure 1. Model of four-grid polarizer (left) and its return losses (dot line – FEM; solid line – impedance method) (right).

Table 1. Final polarizer parameters.

Parameter	*L1*	*L2*	*L3*	*G1*	*G2*	*G3*	*G4*
Dimension (mm)	5.8	5.8	5.8	0.1	0.1	0.1	0.1

The calculated frequency responses of return losses of the polarizer in a case of normal incidence and for an impedance method are shown (the right part of figure 1). It is visible that in frequency range 6 to 20 GHz the return loss entered by the polarizer less than minus 20 dB. To compare in the right part of figure 1 the dot line shows the return losses calculated for the same model of the polarizer by the finite-element method. Coincidence of the compared results received by two different methods confirms reliability of the developed model of the polarizer.

3. The analysis of angle characteristics of the polarizer when scanning

The angle characteristics of four-grid polarizer model were calculated with step of 250 MHz. The analysis showed that with increase of angle of incidence of electromagnetic wave in E-plane, the range of its operating frequencies is displaced in more high-frequency domain (the left part of figure 2). The level of the return loss remains low.

In the same sector of angles of incidence, but in H-plane, the shift of characteristic grows with increase of angles of incidence (up to 4.5 GHz for angle of incidence of 45°; the right part of figure 2). The level of the return loss with increase of angles of incidence until 60° in H-plane grows to 6 dB (at frequency of 6 GHz).

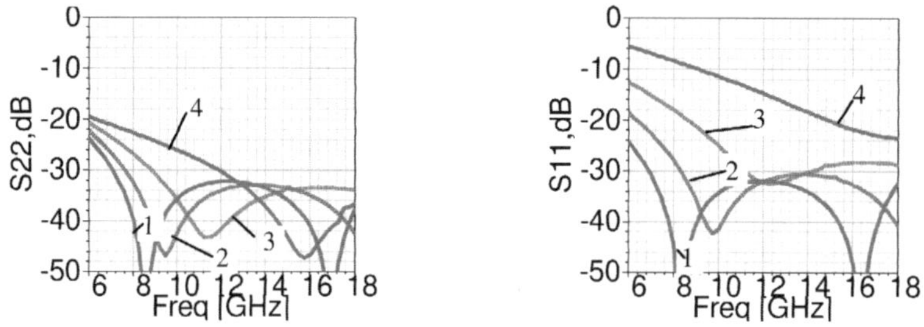

Figure 2. Return losses of the polarizer in E-plane (left) and H-plane (right) in a case of different angles of incidence: 1 - 0°, 2 - 30°, 3 - 45°, 4 - 60°.

Therefore, the designing of a broadband polarizer to operate in a wide sector of scan angles, can consist of two stages. At the first stage the designing of the polarizer in a case of normal incidence is performed by means of an impedance method. Polarizer has to provide the given characteristics in the more wide frequency range. The expansion of the range is carried out towards the lower frequencies. It equals to the shift of frequency response of the polarizer for the angle of incidence at on boundary of scanning sector. At the second stage, a design of a polarizer is optimized to operate in antenna array aperture in the given sector of scan angles.

4. The analysis of antenna array characteristics with the polarizer in the aperture

Effect of the polarizer on array characteristics (VSWR, gain, radiation pattern) is calculated by FEM in the given sector of scan angles. The model of the antenna array consists from linear array of 50 parallel-plate waveguides with the sizes of total aperture of 250 mm × 50 mm. The waveguide aperture size in E-plane was 5 mm; in H-plane – 50 mm. The four-grid model of the 45° rotator of polarization (described above) is optimized for operation in frequency range 8 to 18 GHz in scanning sector ± 45°. Polarizer is placed in front of apertures of waveguides on distance of 10 mm. In figure 3 the antenna array unit cell with the polarizer in an aperture is shown.

Calculations were carried out at frequency of 18 GHz with step of 250 MHz; its precision after 6th steps, when the construction was divided on 4574 tetrahedra, was 0.0007. The VSWR versus frequency for the array with the polarizer in the aperture in a case of different angles of scanning were presented in the left part of figure 4. The VSWR versus angle of on an input of radiators (as a part of antenna array) for scanning in E-plane of waveguide at frequencies 8 GHz, 13 GHz and 18 GHz are given in the right part of figure 4.

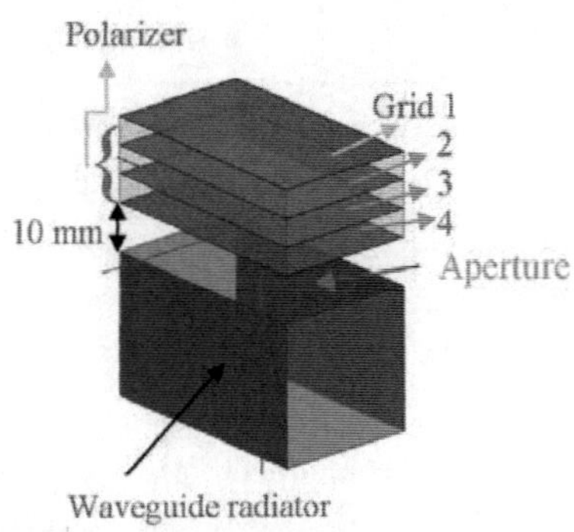

Figure 3. Antenna array unit cell with the polarizer in the aperture.

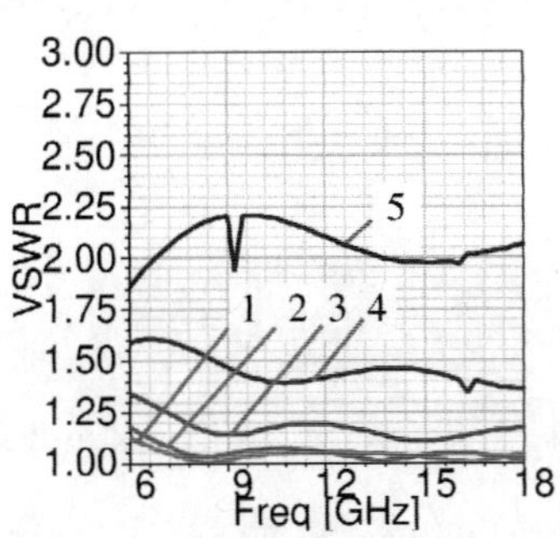

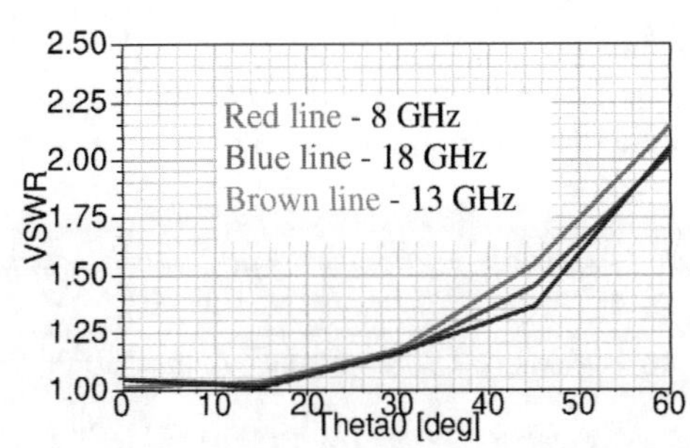

Figure 4. VSWR frequency dependences of the array with the polarizer in the aperture in a case of different angles of scanning: 1 - 0°, 2 - 15°, 3 - 30°, 4 - 45°, 5 - 60° (left) and VSWR angular dependences when scanning in E-plane (right).

The calculations show that in E-plane in the sector of scan angles of ±45° the VSWR on an input of antenna array with the polarizer in the aperture remains not higher than 1.56; VSWR increases to 2.12 in the sector of scan angles of ±60°. On boundary of the sector of scanning of 45° lowering of antenna array gain about 1.5 dB and increase of the main-lobe beamwidth (figure 5) are watched.

Patterns for orthogonal components of far field of the antenna arrays with the polarizer are provided in a case of different scan angles of 0°, 30°, 45° and 60° in E-plane of waveguides at the upper frequency of operating range (figure 5). It is visible that the difference between magnitudes of two orthogonal components of far field of the antenna array grows with increase of the scan angle. For example, it equals about of 3 dB for scan angle of 45°. Therefore the slope angle of E-vector of far field changes with increase of the scan angle.

Also the patterns for far field ellipticity were calculated in a case of the same scan angles. The polarization remains the almost linear in the sector of scan angles up to 60° with ellipticity not worse than 40 dB.

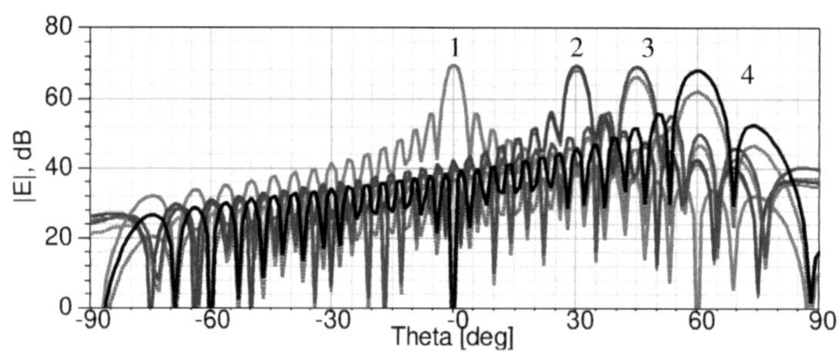

Figure 5. Patterns on orthogonal line polarizations (vertical polarization – solid lines; horizontal polarization – dash lines) at the upper operating frequency in the plane of scanning in a case of different scan angles: 1 - 0°, 2 - 30°, 3 - 45°, 4 - 60°.

5. Conclusion

In the paper we have investigated the operation of broadband polarizer in the aperture of antenna array is analyzed. The effect of the polarizer on the characteristics of the linear array of parallel-plate waveguides in scan sector ±45° in E-plane has been evaluated. Expected VSWR of the element of phased array with the polarizer is less than 1.6 in the frequency range 8...18 GHz. The gain decrease of antenna array about 1.5 dB as well as the increase of the main-lobe beamwidth and the difference between magnitudes of orthogonal components of the far field about 3 dB on boundary of scan sector 45° are watched.

Acknowledgement

This work was performed as project part of the state task No. 8.2461.2014/K of the Ministry of Education of Russia.

References

[1] Cornbleet S 1976 *Microwave Optics. The Optics of Microwave Antenna Design* (Academic Press)
[2] Gimeno B, Cruz J L, Navarro E A and Such V 1994 IEEE Trans. on Antennas and Propag. **42** 912
[3] Amitay N and Saleh A M 1983 IEEE Trans. on Antennas and Propag. **31** 73
[4] Torres R P and Catedra M F 1993 IEEE Trans. on Antennas and Propag. **41** 208
[5] Mallahzadeh A R, Dastranj A A and Hassani H R 2008 Progress in Electromag. Research **B** 67
[6] Semenikhina D V, Klimov A V, Semenikhin A I and Yukhanov Y V 2015 *57th Int. Symp. ELMAR-2015* (Zadar: Croatia) pp 149-152

Conductor disc used to suppress spurious mode and enhance electric coupling in a dielectric loaded combline resonator

T M Pholele[1] and J M Chuma[2]

[1] Department of Basic Sciences, Botswana College of Agriculture, Private Bag 0027, Gaborone, Botswana
[2] Department of Electrical, Electronic, Computer and Telecommunication Engineering, Botswana International University of Science and Technology, Private Bag 0016, Phalapye, Botswana

tpholele@bca.bw

Abstract. The effects of conductor disc in a dielectric loaded combline resonator on its spurious performance, unloaded quality factor (Q_u), and coupling coefficients are analysed using a commercial electromagnetic software package CST Microwave Studio (CST MWS). The disc improves the spurious free band but simultaneously deteriorates the Q_u. The presence of the disc substantially improves the electric coupling by a factor of 1.891 for an aperture opening of 12 mm, while it has insignificant effect on the magnetic coupling.

1. Introduction

Today telecommunication systems operate under strict requirements of high selectivity, low loss, high rejection, wide spurious free stop-bands, small size, low cost and temperature stability [1]. Dielectric loaded combline filters have in the recent past received great attention due to the fact that they offer high Q_u. Their drawback has however been their relatively poor spurious performance. The presence of the dielectric puck significantly reduces the size of the filter.

The spurious free performance of dielectric loaded filters have been improved by among other things, choosing the correct aspect ratio (ratio of diameter to the height of the dielectric) [2], using sandwiched conductor dielectric resonators [3], drilling a hole through the dielectric (ring resonators) [4], correctly positioning the dielectric in the cavity and using conductor loaded dielectric cavity resonators [5]. On the other hand spurious free performance of combline filters has been improved by using stepped impedance resonators [6]. There is a need to design dielectric loaded combline resonators with improved spurious free band.

Coupling is very important in building filters as it is through it that we control the filter bandwidth and selectivity. The selectivity of the filter mainly depends on the position of transmission zeros. These transmission zeros are brought about by cross couplings through apertures or coupling through non resonating modes [7]. Electric coupling (resonators interacting through electric fields) is generally weak compared to magnetic coupling hence there is a need to enhance it (electric coupling).

CST MWS was used to investigate the effect of loading a conductor disc on top of both the post and the dielectric in a dielectric loaded combline resonator on the spurious free performance, Q_u and both the electric and magnetic coupling. The spurious free band improves as the diameter and height of the conductor disc increases but simultaneously deteriorates the Q_u. On the other hand, the presence of the conductor disc improves the electric coupling by 1.891 for an aperture opening of 12 mm, while it has insignificant effect on the magnetic coupling.

2. The effect of the diameter of the disc on the spurious performance and Q_u

Figure 1 shows a resonator which consists of a rectangular enclosure, a cylindrical ring made of high-Q

IOP Conf. Series: Materials Science and Engineering **120** (2016) 012017 doi:10.1088/1757-899X/120/1/012017

ceramic with dielectric constant $\varepsilon_0 = 37$, a centre post and a conductor disc. The dimensions of each resonator are b = 20 mm, L = 23 mm. l =20mm, d = 12.70 mm, d_r = 18 mm, aluminum disc thickness t = 1.6 mm. A dielectric loaded combline resonator filter was initially designed without a conductor disc on top of both the centre conductor post and the cylindrical ring puck. The dielectric loaded combline resonator was then simulated in CST MWS which has been proven to provide accurate results [8]. The resonant frequency and the first spurious mode were noted as 1.758 GHz and 3.698 GHz respectively.

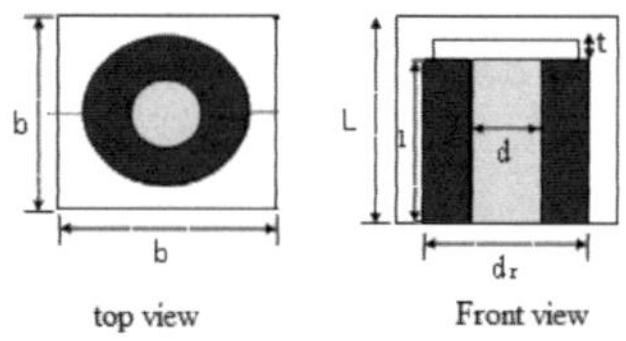

Figure 1. Dielectric loaded combline resonator with a conductor disc on top.

A conductor disc of thickness t was then introduced on top of both the centre conductor and the cylindrical dielectric ring puck (figure 1 right) and the simulations in CST MWS run. In this part of the investigation, the height of the disc was kept the same at 1.6 mm, while the diameter of the disc was varied. The height L, of the enclosure was increased and kept at 24.6 mm so that the effect of the space between the top of the disc and the top of the enclosure remained the same. In this process the resonant frequencies of the modes, the field configurations and the quality factors were observed. It was observed that as the diameter of the disc increases, the frequency of the first spurious mode is driven up in frequency and the resonant mode has its frequency pushed down. The presence of the disc results in reconfiguration of the modes as the electric field that was initially there may not deeply penetrate the conducting disc. This leads to the spurious free region of the resonator improving as the diameter of the disc is increased all the way up to the edge of the dielectric before briefly becoming constant, as indicated in figure 2.

The effect of the diameter of the conductor disc on the Q_u of the resonator was simulated in the CST MWS and the Q_u was found to consistently deteriorate as the diameter increased, as shown in figure 2. Increased surface area of the disc means more resistive losses due to conduction current. A change of the diameter of the disc form 14 mm to 18 mm leads to the spurious free band improving by 109 MHz, and the Q_u deteriorating by 76.3. This therefore calls for a trade-off between obtaining better spurious free performance and reduced Q_u.

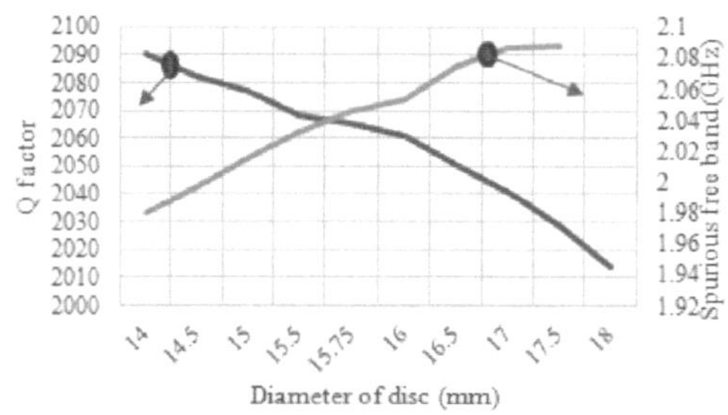

Figure 2. Q factor and spurious free band vary with diameter of the disc.

3. The effect of the height of the conducting disc on the spurious performance and Q_u

The configuration in figure 1 was used in this part of the investigation, but the diameter of the conducting disc was kept constant at 15.75 mm (a diameter which is almost half way between 12.7 mm and 18 mm, the diameter of the dielectric). At this diameter of 15.75 mm we get a spurious free band of 2.045 GHz, an improvement by 132 MHz on the spurious free band of the dielectric loaded combline resonator filter without the conducting disc, and reduction of the Q_u by 32.2 due to the introduction of the conducting disc. The height of the conducting disc was increased and the space between the disc and the top cavity wall was kept the same at 3 mm, that is, every time the height of the disc was increased, the height of the cavity was also increased so that the gap between the disc and the cavity top wall stayed the same.

The simulations in CST MWS revealed that the spurious free band improves as the height of the disc increases. As the height of the conducting disc increases, the frequency of the first spurious mode remains almost the same at around 3.755 GHz (height of 0.2 mm to 1.4 mm) and then drops slightly from 3.753 GHz for a disc height of 1.8 mm to 3.749 GHz for a height of 2.6 mm. On the other hand, the resonant frequency of the resonator gradually reduces from 1.745 GHz for a disc height of 0.2 mm to 1.676 GHz for a disc height of 2.6 mm. This leads to the spurious free band increasing as the height of the disc increases as shown in figure 3.

The Q_u deteriorate as the height of the conducting disc is increased. This relationship is shown in figure 3. A change of the height of the disc from 0.2 mm to 2.6 mm improves the spurious free band by 64 MHz and reduces Q_u by 27.5, which calls for a trade-off.

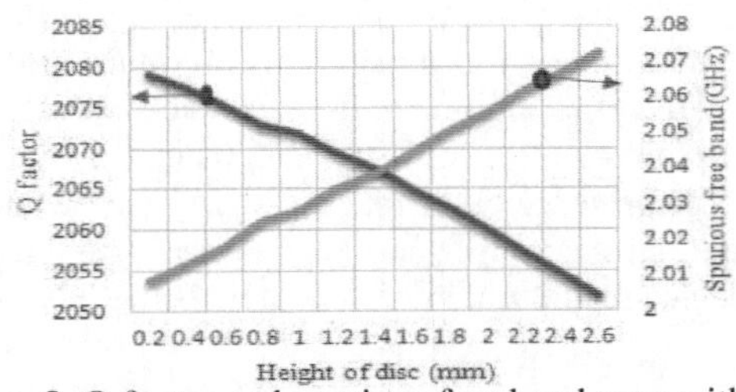

Figure 3. Q factor and spurious free band vary with height of the disc.

4. Effect of the disc on the coupling coefficients

A cylindrical conductor disc of diameter of 15.75 mm and height of 1.6 mm in figure 1 was used in the study. To get f_e and f_m (the resonant frequencies when the symmetry walls were assigned electric and magnetic respectively) two coupled identical combline resonators were built and the symmetry option in boundary conditions was used to assign the common wall as electric and then magnetic. Apertures were milled on the common wall and in each case f_e and f_m noted and coupling coefficient, k computed. Coupling coefficient for identical resonators can be calculated using equation (1) [9]:

$$k = \frac{f_e^2 - f_m^2}{f_e^2 + f_m^2} \tag{1}$$

The disc has minimum effect on magnetic coupling since its presence mainly affects the electric field at the capacitive end of the structure. An aperture milled from the short circuit end of the resonator, for both cases when the disc is loaded and when it is not loaded, show similar results indicating that the disc has minimum effect on magnetic coupling, see figure 4. On the other hand, the electric coupling (aperture milled from the capacitive end of the common wall) of the resonator is significantly affected, particularly from aperture opening from 8 mm to 14 mm, as shown in figure 4. This capacitive aperture opening enhances electric coupling, particularly between aperture openings of 10 mm to 14 mm. A 10 mm aperture opening enhances coupling by a factor of 1.677, while 12 mm aperture opening results in electric coupling being improved by a factor of 1.891. The presence of the disc provides greater surface on which the electric field can begin and end.

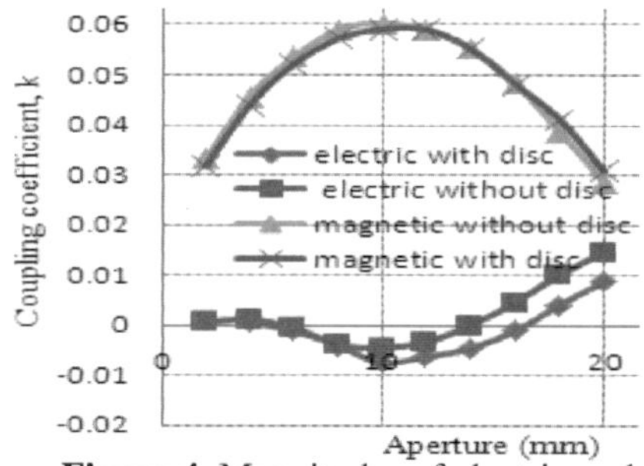

Figure 4. Magnitudes of electric and magnetic coupling coefficients against aperture opening.

Conclusion

It has been shown that the spurious free band of the resonator improves as the height of the loaded disc increases. A change of the height of the disc from 0.2 mm to 2.6 mm improves the spurious free band by 64 MHz and reduces Q_u by 27.5. Similarly, increasing the diameter of the disc also leads to the spurious free band increasing. A change of the diameter of the disc form 14 mm to 18 mm leads to the spurious free band improving by 109 MHz, and the Q_u deteriorating by 76.3 Since increasing the both the height and the diameter of the disc increases the spurious free band and simultaneously deteriorates the Q_u, there is need for a trade-off among the three. The presence of the disc enhances electric coupling, while it has minimum effect on the magnetic coupling. A 10 mm aperture opening enhances electric coupling by a factor of 1.677, while 12 mm aperture opening results in electric coupling being improved by a factor of 1.891.

References

[1] Chuma J M and Mirshekar-Syahkal D 2000 Compact dielectric loaded combline filters with low insertion loss Microw. Conf. pp 1-4

[2] Kazemi M J Dadash M S and Safian R 2011 Design and implementation of a bandpass filter using dielectric resonators with new excitation structure IET Microw. Antennas Propag. **5**(12) pp 1416-1423

[3] Weily A R and Mohan A S 2001 Microwave filters with improved spurious performance based on sandwiched conductor dielectric resonator IEEE Trans. Microwave Theory Tech. **49**(8) pp 1501-1509

[4] Chen J W and Zaki K A 1991 Dielectric ring resonators loaded in waveguide and on substrate IEEE Trans. Microwave Theory Tech. **39**(12) pp 2069-2076

[5] Hunter I C Dassonville V and Rhodes J D 1999 Dual-Mode filters with conductor loaded resonators IEEE Trans. Microwave Theory Tech. **47**(12) pp 2304-2310

[6] Chen H H et al 2011 Investigations on improving spurious performance of a coaxial combline Filter IET Microw. Antennas Propag. **5**(4) pp 459-467

[7] Lapidus A D and Rossiter C 2002 Cross-coupling in microwave bandpass filters Microw. J.

[8] Pholele T M and Chuma J M 2014 Effects of combined apertures on the magnitudes of coupling coefficients of combline resonators IOP Conf. Ser.: Matter. Sci. Eng. **67**

[9] Zaki K A and ChenC 1987 Coupling of non-axially symmetric hybrid modes in dielectric resonators IEEE Trans. Microwave Theory Tech. **35**(12) pp 1136-1142

IOP Conf. Series: Materials Science and Engineering**120** (2016) 012018 doi:10.1088/1757-899X/120/1/01201ξ

Impact of electromagnetic fields on human vestibular system and standing balance: pilot results and future developments

A Allen[1,4], S Villard[1,2], M Corbacio[1,2], D Goulet[5], M Plante[5], M Souques[6], F Deschamps[7], G Ostiguy[5], J Lambrozo[6], A W Thomas[1,3] and A Legros[1,2,3,4,8]

[1] Human Threshold Research Group - Bioelectromagnetics, Imaging Program, Lawson Health Research Institute, London, ON, N6A 4V2, Canada
[2] Department of Medical Biophysics, Western University, London, ON, N6A 3K7, Canada
[3] Department of Medical Imaging, Western University, London, ON, N6A 3K7, Canada
[4] School of Kinesiology, Western University, London, ON, N6A 3K7, Canada
[5] Hydro-Québec, Montreal, QC, H2Z 1A4, Canada
[6] Service des études médicales, EDF, Paris, France
[7] Service Environnement Réseaux, RTE, Paris, France

Abstract. Although studies have found that extremely low-frequency (ELF, < 300 Hz) magnetic fields (MF) can modulate human standing balance, the acute effects of electromagnetic fields on standing balance have not been systematically investigated. This work aims to establish the threshold for acute standing balance modulation during ELFMF exposure. One hundred volunteers will be exposed to transcranial electric stimulations (Direct Current - DC and Alternating Current - AC, 1 mA) and ELFMF (0 to 160 Hz, 0 to 100 mT). The displacement of their center of pressure will be collected and analyzed as an indicator of vestibular performance. During pilot testing (n=6), we found increased lateral sway with DC, and to a lesser extent, AC exposure. The ELFMF exposure system still needs to be adapted to allow meaningful results. Future protocol design will test for possible effects due to exposures in the radiofrequency range (i.e. above 3 kHz). These results will contribute to the literature documenting exposure guidelines aiming to protect workers and the general public.

1. Introduction

Power frequency magnetic fields (MF – 50 and 60 Hz) result from electricity generation and distribution and from the use of electrical household appliances. MF levels up to 2mT can occur near certain household appliances [1, 2]. These time-varying MF induce electric fields and currents in conductors, including exposed biological systems. From a public health perspective, it is crucial to know the thresholds at which ELFMF and the corresponding induced internal electric fields and currents have an acute biological effect. More specifically, at what exposure threshold does the MF acutely modulate neurophysiological processes

[8] To whom any correspondence should be addressed.

in humans? International agencies, such as the Institute of Electrical and Electronics Engineers (IEEE) and the International Commission on Non-Ionizing Radiation Protection (ICNIRP), provide exposure recommendations and guidelines protecting the public and workers from MF exposure and its potential adverse effects resulting from induced electric fields [3, 4]. Human balance control, and its modulation under EMF exposure conditions, is one of the meaningful neurophysiological outcomes reviewed in these guidelines. Indeed, standing balance is shown to be particularly responsive to Extremely Low Frequency (ELF, < 300 Hz) MF exposure [5, 6, 7]. Interestingly, standing balance is a complex process involving vestibular, visual, and sensory-motor signals. Glover et al. hypothesized that observed sway pattern perturbations were due to forces on the vestibular organ and induced currents in the vestibular nerve [8]. Nevertheless, these previous studies used low MF exposure levels (below 1 mT) and did not focus on establishing an acute threshold response to higher (up to 100mT) MF exposure levels.

Acute vestibular perturbations translating to a loss of balance in humans have, however, been well documented for a direct current stimulation (DC) called galvanic vestibular stimulation (GVS). Indeed, GVS at 1mA leads to an acute standing balance effect [9]. Interestingly, it has been reported that a 60Hz MF stimulation at 100mT should induce current intensity of such order [10, 11]. Our first hypothesis is that a 60Hz MF stimulation at 100mT will acutely impact standing balance in a similar fashion to a 1mA DC exposure with GVS. Therefore, the main goal of this study is to find the threshold for an acute standing balance response somewhere between 0 and 100mT (tested in 10mT increments).

The most documented physiological biomarker of ELFMF stimulation is the magnetophospene, a flickering visual perception. Magnetophosphene perception serves as a threshold detection reference for IEEE recommendations and ICNIRP guidelines. Research has shown that magnetophosphenes are more easily detected at 20Hz [12]. This finding suggests that magnetic flux density threshold for an acute response might be dependent on the frequency of the MF stimulation. Therefore, the second objective of this study is to evaluate the effect of different frequencies of ELFMF (20Hz, 60Hz, 90Hz, 120Hz, and 160Hz). Since changing the MF stimulation frequency for a given flux density also changes the induced current intensity, we also tested the same frequencies with a fixed 1mA transcranial Alternating Current Stimulation (tACS). tACS consists of externally applying an alternating current and has been used in order to obtain a functional or even a clinical effect. Specifically, it has been shown to reduce resting tremor in Parkinson's patients, however vestibular effects have not yet been explored [13].

2. Methods

2.1. Subjects

One hundred healthy volunteers (aged 18-55) will be recruited and tested. Exclusion criteria include history of vestibular-related pathology or dysfunction, chronic illnesses (e.g., cardiovascular diseases such as hypertension, ischemia, and cerebrovascular disease) and neurological diseases that affect normal body movement (e.g., Parkinson's disease or Multiple Sclerosis). Participants will have their eyes closed during the data collection portion of the protocol and so participants with visual impairments are eligible.

2.2. Experimental Devices

We will be using a force plate (OR6-7-1000, AMTI, USA) recording postural sway (displacement of the Center Of Pressure, COP), using an A/D module (NI USB-6251), driven by LabView 14.0.1 (National Instruments, USA). Galvanic vestibular stimulation (DC stimulation, 1 mA) and tACS (AC stimulation, 1 mA, 20-160 Hz) will be delivered using the StarStim system (Neuroelectrics, Spain). Exposure will be directed at the mastoid level in order to target the vestibular system. MF exposure will be delivered via a headset exposure system (two 375 turn-coils of 5.2 cm diameter, with a 2 cm diameter laminated core of permendur-49 – The Goodfellow Group, Coraopolis, PA, USA). The exposure will be delivered behind the ear (just above the mastoid process of the temporal bone), targeting the vestibular system, located 2.5 cm deep from the external auditory canal.

2.3. Experimental Design
Five groups of 20 participants will be tested. Each group will be assigned to a specific stimulation frequency: 20Hz, 60Hz, 90Hz, 120Hz, and 160Hz. Participants will be standing (feet together, eyes closed) on a force plate covered in a foam layer to maximize vestibular contribution and will be exposed to three stimulation modalities: direct current (DC), tACS and MF.

Each exposure will last 5 seconds followed by a 30-second rest and will be repeated on the right and left side of the head. Thus, for each group, DC stimulations will consist of 2 exposures (Right and Left), tACS at the assigned frequency will consist of 2 exposures (Right and Left), and MF stimulation at the assigned frequency will consist of 22 exposures (0 to 100mT in 10mT increments for Right and Left). Exposures will be presented in a random order.

This protocol was approved by Western University's Health Sciences Research Ethics Board (#106122).

3. Results
Transverse and sagittal mean sway (cm), sway velocity (X and Y, cm/s), sway path (cm), and sway area (cm^2), calculated from COP data will be used for statistical analysis [6, 14]. The statistical analysis will consist of a within-subjects ANOVA (flux densities x side of exposure) with a between-subjects factor (frequency groups). Six pilot subjects have been tested so far. Based on preliminary inspection of these data, DC stimulation seemed to affect sway patterns and demonstrated a clear responsive tilt towards the side of the head containing the stimulation anode (Fig. 1 and 2). Similar acute effects appear to result from tACS, however, the MF condition (pilot tested at a single flux density of 100mT) showed minimal effects on standing balance (Fig. 2). This was expected since the pilot exposure device was stationary and therefore could not allow for much participant movement while maintaining exposure. The use of an ambulatory unilateral exposure device currently being developed should improve this result.

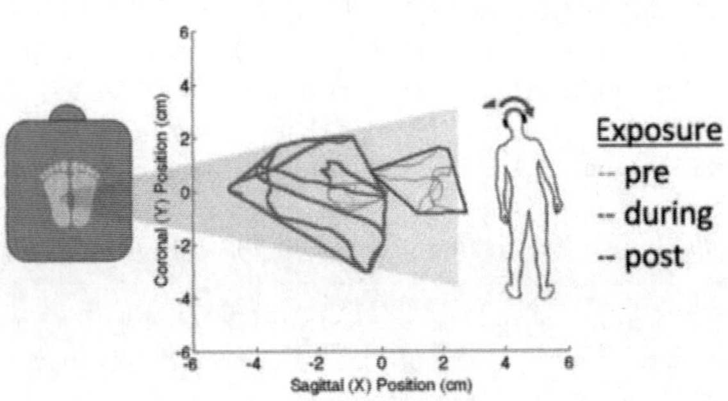

Figure 1. Horizontal sway patterns from left 5s GVS exposure in red (1mA, 1s ramp), 5s pre exposure period (light blue) and 5s post (dark blue).

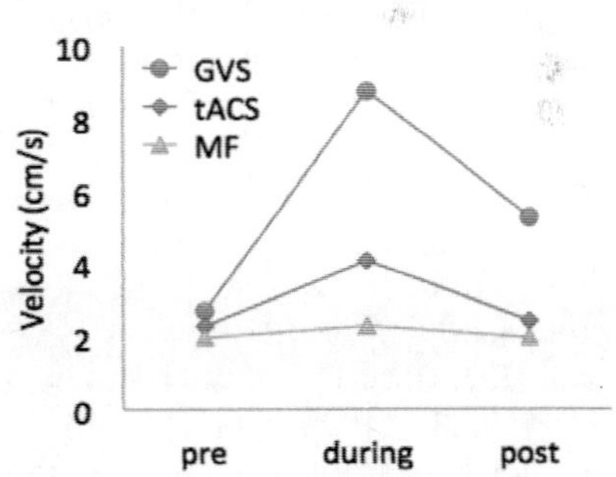

Figure 2. Subject average sway velocity of oscillation for 5s right exposure to GVS (1mA), tACS (1mA, 120Hz) and MF (120Hz, 100mT).

4. Discussion
The main aim of this study is to establish a flux density threshold at which acute postural responses are detected for a 60 Hz MF exposure. The second aim of this study is to identify how this threshold is influenced by the frequency of the stimulation.

So far, with six pilot subjects tested, data appears to be promising in demonstrating an acute effect of DC and tACS on human standing balance. Higher frequencies appeared to show a greater effect in the tACS condition. Regarding the MF condition, different flux density effects have not been tested yet, however the future protocol will use a newly developed MF exposure device allowing for free movement of the participants and therefore a more thorough exploration of the MF effects on standing balance. Data from this study will contribute to the literature supporting the rationale for exposure guidelines protecting public and worker safety, developed by international agencies, such as ICNIRP and IEEE.

Considered future developments could question the possible direct impact of radiofrequencies on the vestibular system via electrophysiological mechanisms (in the kHz range) or via a thermal effect (at cell phone frequencies - 900MHz and above). A thermal effect could indeed trigger a possible vestibular caloric response, which is described as a rapid eye movement due to an acute temperature change in the external auditory canal [15]. The vestibular system could therefore be an interesting biological target adapted to test for a thermal effect in humans.

5. References

[1] Gauger J R 1985 Household appliance magnetic field survey *Power Eng.* **5** 2436

[2] Gandhi O P, Kang G, Wu D and Lazzi G 2001 Currents induced in anatomic models of the human for uniform and nonuniform power frequency magnetic fields *Bioelectromagnetics.* **22** 112

[3] ICNIRP 1998 Guidelines for limiting exposure to time-varying electric, magnetic, and electromagnetic fields (up to 300 GHz) *Health Phys.* **74** 494

[4] IEEE 2002 *IEEE Standard for safety levels with respect to human exposure to electromagnetic fields, 0-3 kHz* (New York: IEEE)

[5] Legros A, Corbacio M, Beuter A, Modolo J, Goulet D, Prato F S and Thomas A W 2012 Neurophysiological and behavioral effects of a 60 Hz, 1,800 muT magnetic field in humans *Eur J Appl Physiol.* **112** 1751

[6] Thomas A W, Drost D J and Prato F S 2001 Human subjects exposed to a specific pulsed (200 microT) magnetic field: effects on normal standing balance *Neurosci Lett.* **297** 121

[7] van Nierop L E, Slottje P, Kingma H and Kromhout H 2013 MRI-related static magnetic stray fields and postural body sway: a double-blind randomized crossover study *Magn Reson Med.* **70** 232

[8] Glover P M, Cavin I, Qian W, Bowtell R and Gowland P A 2007 Magnetic-field-induced vertigo: A theoretical and experimental investigation *Bioelectromagnetics.* **28** 349

[9] Fitzpatrick R C and Day B L 2004 Probing the human vestibular system with galvanic stimulation *J Appl Physiol.* **96** 2301

[10] Merlet I, Birot G, Salvador R, Ardekani B, Mekonnen A, Frish A, Ruffini G, Miranda P C and Wendling F 2013 From oscillatory transcranial current stimulation to scalp EEG changes: a biophysical and physiological modeling study *PLoS One.* **8** e57330

[11] Salvador R, Ramirez F, V'yacheslavovna M and Miranda P C 2012 Effects of tissue dielectric properties on the electric field induced in tDCS: a sensitivity analysis *Conf Proc IEEE Eng Med Biol Soc.* 787

[12] Lovsund P, Oberg P A, Nilsson S E and Reuter T 1980 Magnetophosphenes: a quantitative analysis of thresholds *Med Biol Eng Comput.* **18** 326

[13] Brittain J S, Probert-Smith P, Aziz T Z and Brown P 2013 Tremor suppression by rhythmic transcranial current stimulation *Curr Biol.* **23** 436

[14] Despres C, Lamoureux D and Beuter A 2000 Standardization of a neuromotor test battery: the CATSYS system *Neurotoxicology.* **21** 725

[15] Hain T C 2012 Vestibular testing. American Hearing Research Foundation

AUTHOR INDEX

Abdulla, P 25
Alicalapa, F 29
Allen, A 69
Andriambeloson, J. A. 37
Beeharry, Girish Kumar 45
Chakraborty, Tamal 5
Chuma, J M 65
Çiftçi, T 17
Corbacio, M 69
Delamare, A. 21
Delaroche, F 21
Deschamps, F 69
Douyère, A 29
Ergül, Ö 17
Galande, Shridhar 33, 53
Goulet, D 69
Hartmann, Markus 13
Heuberger, Albert 13
Jamadar, Sachin 33, 53
Jasmine, P M 25
Joshi, B. C. 33, 53
Kandukuri, Somasekhar 9
Karaosmanoglu, B 17
Kate, G. 33, 53
Knyazev, S. 41
Korotkov, A. 41
Lalléchère, Sébastien 57
Lambrozo, J 69
Latry, O. 21
Lebreton, J M 1
Legros, A 69
Lorion, R 1
Lorion, Richard 9
Luk, J-D Lan Sun 29
Mahajan, S. P. 33, 53
Manoharan, P. K. 33, 53
Misra, Iti Saha 5
Murad, N M 1
Murad, Nour 9
Naidu, Arun 53
Normand, A. 21
Nowak, Thorsten 13
Ostiguy, G 69
Panchenko, B. 41
Patil, R. A. 33, 53
Patino-Studencki, Lucila 13
Pethe, Kaiwalya 33, 53
Pholele, T M 65
Plante, M 69
Prayag, Vinand 45
Raphika, P M 25
Ravelo, B. 21
Ravelo, Blaise 57
Rivière, J 29
Robert, Jörg 13
Roy, Jayashree 33, 53
Semenikhin, A I 61
Semenikhina, D V 61
Shabunin, S. 41
Souques, M 69
Thakur, Atul 57
Thielecke, Jörn 13
Thomas, A W 69
Villard, S 69
Vurpillot, F. 21
Vydelingum, Nazir 45
Wiid, P. G. 37
Winberg, Simon 49
Yukhanov, Y V 61
Zhao, L. 21